# Lecture Notes in Computer Science 15657

Founding Editors

Gerhard Goos

Juris Hartmanis

The series Lecture Notes in Computer Science (LNCS), including its subseries Lecture Notes in Artificial Intelligence (LNAI) and Lecture Notes in Bioinformatics (LNBI), has established itself as a medium for the publication of new developments in computer science and information technology research, teaching, and education.

LNCS enjoys close cooperation with the computer science R & D community, the series counts many renowned academics among its volume editors and paper authors, and collaborates with prestigious societies. Its mission is to serve this international community by providing an invaluable service, mainly focused on the publication of conference and workshop proceedings and postproceedings. LNCS commenced publication in 1973.

Laura Carnevali · Josu Doncel
Editors

# Computer Performance Engineering

21st European Workshop on Performance Engineering, EPEW 2025
Catania, Italy, June 26, 2025
Revised Selected Papers

 Springer

*Editors*
Laura Carnevali 
University of Florence
Florence, Italy

Josu Doncel 
University of the Basque Country
Leioa, Spain

ISSN 0302-9743     ISSN 1611-3349 (electronic)
Lecture Notes in Computer Science
ISBN 978-3-032-16344-8     ISBN 978-3-032-16345-5 (eBook)
https://doi.org/10.1007/978-3-032-16345-5



This Springer imprint is published by the registered company Springer Nature Switzerland AG
The registered company address is: Gewerbestrasse 11, 6330 Cham, Switzerland

If disposing of this product, please recycle the paper.

# Preface

The European Performance Engineering Workshop (EPEW) aims to bring together researchers working on performance evaluation of systems from theoretical and practical viewpoints. The concept of performance is considered in its broadest sense including the notions of Quality of Service, scalability, reliability, availability and systems management, among others.

The 21st edition of EPEW was held on the 26th of June 2025 in Catania (Italy), co-located with the 39th ECMS International Conference on Modelling and Simulation. This edition was a very fruitful workshop with numerous presentations dealing with performance evaluation research problems.

We acknowledge the organizers of ECMS for their help in the organization of this edition of EPEW. We also thank Reinhard German for giving an excellent keynote talk in this edition of EPEW, which was entitled "Analyzing Energy Systems with Agent-Based and Data-Centric Distributed Simulation". Furthermore, we acknowledge the work of the Program Committee, which was formed by 37 top-level researchers from different countries.

In this edition, there was a two-phase review process. First, we solicited short papers of at most 7 pages for presentation at the conference; these short papers will not be published. Then, authors of accepted short papers were invited to submit a full paper. Accepted full papers appear in this volume. Initially, we received 15 short papers, from which 12 were accepted for presentation at the conference. Besides, we received 11 full paper submissions and all of them were accepted for publication in this volume. Each paper was single-blindly reviewed in each of the rounds by at least three reviewers.

Finally, we would like to thank all the authors that contributed to this workshop, for either submitting a paper to the workshop or giving a talk during the event, or submitting a full paper for publication in this volume.

November 2025

Laura Carnevali
Josu Doncel

# Organization

## Program Committee Chairs

Laura Carnevali      University of Florence, Italy
Josu Doncel      University of the Basque Country, Spain

## Local Organizer

Marco Scarpa      University of Messina, Italy

## Program Committee

Salvador Alcaraz      Miguel Hernández University of Elche, Spain
Elvio Gilberto Amparore      University of Turin, Italy
Paolo Ballarini      CentraleSupélec, France
Simona Bernardi      University of Zaragoza, Spain
Marco Bernardo      University of Urbino, Italy
Peter Buchholz      TU Dortmund, Germany
Giuliano Casale      Imperial College London, UK
Davide Cerotti      Technical University of Milan, Italy
Carina da Silva      University of Münster, Germany
Dieter Fiems      Ghent University, Belgium
Matthew Forshaw      Newcastle University, UK
Jean-Michel Fourneau      University of Versailles Saint-Quentin-en-Yvelines, France
Pedro Pablo Garrido Abenza      Miguel Hernández University of Elche, Spain
Marco Gribaudo      Technical University of Milan, Italy
Antonio Gomez Corral      Complutense University of Madrid, Spain
Nikolas Herbst      University of Würzburg, Germany
András Horvath      University of Turin, Italy
Mauro Iacono      Università degli Studi della Campania "Luigi Vanvitelli", Italy
Emilio Incerto      IMT School for Advanced Studies Lucca, Italy
Alain Jean-Marie      INRIA, France
Carlos Juiz      University of the Balearic Islands, Spain
Lasse Leskela      Aalto University, Finland

| | |
|---|---|
| Francesco Longo | University of Messina, Italy |
| Diletta Olliaro | Ca' Foscari University of Venice, Italy |
| Marco Paolieri | University of Southern California, USA |
| Tuan Phung-Duc | University of Tsukuba, Japan |
| Agapios Platis | University of the Aegean, Greece |
| Markus Siegle | University of the Bundeswehr Munich, Germany |
| Ioannis Stefanakos | University of York, UK |
| Nigel Thomas | Newcastle University, UK |
| Benny Van Houdt | University of Antwerp, Belgium |
| Marco Vieira | University of North Carolina at Charlotte, USA |
| Joris Walraevens | Ghent University, Belgium |
| Armin Zimmermann | Technische Universitat Ilmenau, Germany |

## Additional Reviewers

Lukas Horn
Manuel Fabiano
Vincenzo Bucaria

# Contents

# Enhancing Multiformalism Models with Rewrite Engines


Enrico Barbierato[1], Lorenzo Capra[2]([✉]), Marco Gribaudo[3], and Mauro Iacono[4]

[1] Università Cattolica del Sacro Cuore, Via Garzetta 48, 25121 Brescia, Italy
`enrico.barbierato@unicatt.it`
[2] Università degli Studi di Milano, via Celoria 18, 20133 Milan, Italy
`lorenzo.capra@unimi.it`
[3] Politecnico di Milano, via Ponzio 34/5, 20133 Milan, Italy
`marco.gribaudo@polimi.it`
[4] Università degli Studi della Campania "L. Vanvitelli", viale Lincoln 5,
81100 Caserta, Italy
`mauro.iacono@unicampania.it`



**Abstract.** This paper explores the enhancement of multiformalism modeling frameworks by integrating rewrite engines, with a focus on SIMTHESys, a framework designed for combining and solving heterogeneous formalisms. Multiformalism modeling enables the use of the most suitable formalism for each subsystem while maintaining an overall coherent representation of the system. As model complexity and the need for dynamic adaptability increase, static modeling structures become inadequate. To address this, we extend SIMTHESys with capabilities for dynamic transformations and late binding by incorporating a rewriting system, specifically leveraging the `Maude` engine. This integration allows models to evolve during their solution processes, supporting more flexible, efficient, and modular analyses. We present a case study involving the dynamic management of a server system modeled through stochastic Petri nets and multiclass queuing networks, demonstrating the feasibility and advantages of the proposed approach. Experimental results highlight how rewriting enhances model expressiveness and efficiency in complex system evaluations.

**Keywords:** Multiformalism modeling · Rewriting systems · Model reconfiguration · Performance evaluation · Petri nets · Queuing networks


## 1 Introduction

Real systems exhibit a nature characterized by the coexistence of different aspects that influence each other. Especially during the design or assessment process of such systems, these aspects have to be modeled in order to allow an evaluation of the system. Modeling implies a simplification aimed at keeping the most important features of a system, but an effective analysis requires



L. Carnevali and J. Doncel (Eds.): EPEW 2025, LNCS 15657, pp. 1–16, 2026.
https://doi.org/10.1007/978-3-032-16345-5_1

sound tools, including representations, which allow a proper understanding of the system and an effective representation of the system.

Representations need languages, which are developed in coherence with the purpose of the representation and the semantic content of the representation: different aspects imply different representations to minimize loss of information, cognitive distance between the user of the model and the contents of the model, and different analysis techniques. Specialists in each aspect define and develop their own tools and their own culture to best suit their goals. The more complex a system, the more the aspects which have to be properly represented: consequently, modeling approaches based on a single modeling formalism either suffer of a great complexity of the formalism, making the learning curve steeper and the immediateness of representations lower, or excessively simplify representations, so that the model is not able to keep a sufficient level of detail about the system. Multiformalism modeling is a technique aiming at allowing for a coherent and coordinated coexistence of different modeling formalisms, defined to allow the most natural representation to each aspect of the system. It is based on complex frameworks that try to move the effort from model users to their internal mechanisms, which generally include analysis tools. The more flexible the framework, also in terms of the possibility of defining custom formalisms and combining them at the user's needs, the more sophisticated the tools have to be and the more computationally complex they are. This produces a need for an equally flexible definition of the analysis tools and for the definition of analysis strategies capable of lowering the computational complexity of the analysis, e.g., by decomposition, composeability, abstract specification of analysis procedures, or symbolic tools, to avoid problems such as submodel misalignment, state space explosion, or low usability. Finding profitable approaches to make complex multiformalism models analyzable requires coordinating several different strategies. In this paper, we show the integration of the graph rewriting feature in SIMTHESys, a multiformalism modeling framework capable of allowing the experimentation of novel modeling formalisms and of generating (suboptimal) multiformalism evaluation tools, to include some features of the workflow-based model evaluation approach adopted in OsMoSys, our previous multiformalism modeling framework. We describe the integration process, the emerging advantages, and the new potential by means of a case study from a previous paper of ours in which SIMTHESys was used for the evaluation: This will allow interested readers to compare the two approaches while keeping this paper self-contained.

The paper is organized as follows: the next Section provides the background on multiformalism modeling and rewriting; Sect. 3 presents related work; Sect. 4 introduces the reader to multiformalism modeling solution process strategies; Sect. 5 focuses on SIMTHESys extensions for rewriting; Sect. 6 details the case study; Sect. 7 presents `Maude` and its use within SIMTHESys applied to the analysis of the case study. Finally, Sect. 8 closes the paper.

## 2  Background

Multiformalism modeling is a methodological approach to the generalized performance evaluation of systems, based on integrating multiple modeling techniques and their associated evaluation tools. This enables each aspect of a system—or of its subsystems—to be represented using the most suitable formalism, selected for its appropriateness in terms of expressiveness and interpretability, while preserving a unified and coherent view of the system as a whole. The properties of complex multiformalism models can be assessed through the algorithmic or algebraic composition, potentially dynamic or nested, of results obtained from the independent evaluation of submodels. These results are integrated into global high-level properties that may differ in nature. The evaluation process may proceed through various structured methods, including solution workflows, semantic-preserving transformations of submodels into a more analytically advantageous formalism, and the conditional generation of model components in response to specific parameters or contextual conditions. Furthermore, it is possible to automatically synthesize dedicated evaluation tools based on the semantic descriptions of the elements within the constituent formalisms. Assuming the accuracy and validity of the evaluation methodology, the computational efficiency of the process is influenced by the adopted strategy, the inherent heterogeneity of the involved modeling formalisms, the availability of optimized and composable evaluation tools, and the extent to which external specialized analysis tools can be integrated into the overall evaluation pipeline. In most existing multiformalism modeling frameworks, model composability is provided as a mechanism to support both model design and the management of complexity. Although the specific advantages of composability depend on the features of the individual framework, it generally enables structural modularity within models. A modular approach relies on well-defined composition mechanisms and semantics and often adopts a graph-based organization of model components. This structure facilitates the use of rewriting systems techniques, which enhance analytical flexibility and enable the incorporation of dynamic behaviors, such as redefining parts of a model or entire (sub)models, in response to conditions arising during evaluation. The potential of rewriting systems includes on-the-fly model transformation, dynamic evaluation workflows, and the selection between alternative module definitions or implementations. When effectively integrated with the modeling and evaluation semantics of a framework, rewriting systems can support the construction of dynamic models, enabling more efficient or adaptive evaluation strategies. As models increase in complexity and scale, the need for adaptability during evaluation becomes paramount. Static structures can limit the expressiveness and efficiency of model analysis, particularly in cases where system behavior evolves or where alternative model components should be selected dynamically. To address these challenges, we consider the integration of a rewriting system into the multiformalism modeling process. A rewriting system provides a formal rule-based mechanism for specifying structural or behavioral transformations within a model. This allows the model components, whether individual elements or entire submodels, to be redefined or reconfigured during evaluation, based on context, intermediate results, or optimization criteria. Such dynamic behavior supports features like late binding, where decisions about model composition or implementation are deferred until runtime, and enables adaptive workflows, where model structure evolves in response to system dynamics. In the context of multiformalism modeling, the incorporation of a rewrite engine offers several benefits. It formalizes model evolution, supports modular and reusable designs, and facilitates the construction of dynamic models whose structure may change over time. This increases both modeling flexibility and computational efficiency, especially when managing state space growth or coordinating the evaluation of heterogeneous submodels. This work proposes the integration of rewriting system capabilities into SIMTHESys (Structured Infrastructure for Multiformalism modeling and Testing of Heterogeneous formalisms and Extensions for SYStems), a modeling framework designed to support the combination and analysis of diverse formalisms, the experimentation of novel modeling formalisms, and the generation of (suboptimal) multiformalism evaluation tools, with a focus on the exploitation of the native extensibility of the framework. This approach takes advantage of the `Maude` system as a rewrite engine, allowing the application of transformation rules during model evaluation. The integration empowers SIMTHESys to handle dynamically reconfigurable models and to support new analysis strategies through on-the-fly adaptation and model restructuring.

## 3   Related Work

Multiformalism modeling [1,16,17,23], often coupled with *multisolution*, has been explored in literature in both the case of fixed combination of modeling formalisms (as in DEDS [4], SMART [11] and SHARPE [24]) and in the case of dynamic combinations (AToM$^3$ [20] and Möbius [13]). The work presented in this paper is based on OsMoSys [15] and SIMTHESys [2,19], which adopt a multisolution approach, together with Möbius. Möbius operates by generating an optimized executable model, based on a descriptive model which may include specific model elements behaviors defined programmatically; OsMoSys solves models utilizing the orchestration of a workflow, which activates external independent tools (*solvers*) according to intermediate results of submodels evaluations. In turn, SIMTHESys, which defines models by explicitly separating the description of models and the description of the model elements which are proper of a modeling formalism, exploits this description of formalisms and their elements both to define inter-formalisms interaction mechanisms and to provide (suboptimal) analysis primitives, using both to automatically generate multiformalism *solvers* for all models based on a same combination of formalisms: this natively allows the definition of new formalism and of new formalisms combinations (and one of the purposes of SIMTHESys is to let users experiment with new formalisms, to implement then optimized specific tools if experimentation is productive) with a limited need for additional software development. A synopsis of the main characteristics of these tools can be found in [3]. To understand how rewriting can enhance multiformalism modeling, the following section presents how the solution process is handled in SIMTHESys and how it is extended to support transformations.

## 4    Solution Processes

One of the advantages of the OsMoSys approach to model analysis (in the following, model *solution*, in the jargon of OsMoSys and SIMTHESys) is the possibility of defining models with a dynamic structure. This is obtained because solution happens as an orchestration of applications of solvers to submodels in isolation: a solution process is defined as a workflow which, according to the tree-like compositional structure of a OsMoSys model, manages the solution, starting from terminal submodels, which do not depend on any internal submodel, and applying the solution steps until the overall model is solved and both submodel-wide and model-wide results are obtained. This allows two main advantages: the possibility of solving submodels in isolation and the dynamic transformation of models during the solution process, because parts of the model, according to partial results or conditions which occur during the solution, can be generated or transformed, or can be implemented differently, choosing between alternative implementations, sharing the same interface, of a submodel (this is known as the *late binding* feature of OsMoSys). These possibilities allow to reduce the complexity of the solution by breaking it into partial solutions followed by combinations and by selecting the optimal implementation (for example, this allows to limit the state space explosion problem)[1]. SIMTHESys, instead, analyzes a model, extracts the used formalism, generates a suitable multiformalism solver by choosing between the elementary analysis components (*solving engines*, in the jargon of SIMTHESys) suitable for the chosen formalisms, and solves the overall model by applying the behaviors which characterize each dynamic evolution of each element of each model according to the evolution of the global state of the model and the local state of each element.

While the first approach operates at "solving time" with a process which can be compared to interpretation in programming languages, and the complexity lies in the need for a specification of the needed solution workflow, in the second the process operates in two stages, a first one in which the formalisms used in the model are analyzed to generate, with the support of the SIMTHESys SIMTHE-Sys*ER* tool [18], a multiformalism *solver* for that combination of formalisms, which may be compared to a specialized virtual machine in programming, and a second,in which the model (and any other model using the same combination of formalisms) is solved by the solver, which can be compared to execution of a program compiled for the virtual machine in programming. A SIMTHESys model has a modular structure but is not solved by modules, because of the organization of the solver: large models can be anyway managed by SIMTHE-Sys by exploiting a simulation based solution engine, but its original design was not conceived to support the generation at solution time of parts of the model, which requires the explicit specification of alternative coexisting submodels if needed and less performing solution processes both in analytical solutions and simulation solutions. Introducing OsMoSys-like late binding or implementing

---

[1] Interested readers can find significant examples in [14, 21] and [22].

solving time transformations for submodels in SIMTHESys is a way to improve performance and flexibility, and to simplify model definition.

Concerning other approaches, this solution allows for reducing the complexity of the solution process while preserving both the natural structure of formalism dynamics (e.g., the state space), to keep a fully coherent internal representation, and, in case, enable access to it, and the essential feature of SIMTHESys as a formalism design, definition, and experimentation tool. For example, the solution approach of Möbius [13], probably the most computationally efficient within multiformalism modeling tools, or the solution approach of ORIS [6], probably the most sophisticated within performance-oriented state-space based analysis tools, would not match with the needs of SIMTHESys, as they do not target agile definition of new formalisms or immediate customization of existing formalisms, and have excellently optimized simulation or state-space generation and management approaches aiming at the solution rather than at the process and the exploration of formalisms possibilities.

## 5    Extending SIMTHESys with Transformations and Late Binding

Submodels in SIMTHESys have a semantic purpose and logically delimit self-consistent model parts, possibly having their own state. They allow reuse and composition, but do not play any specific role in the solution process. Nevertheless, they enable defining SIMTHESys models as hierarchical graphs, which offers the possibility of applying rewriting systems advantages to introduce some of the OsMoSys features seen (namely, dynamic transformations and late binding) and improve the performances of solution processes based on the state space. As rewriting is a novel feature that affects the structure of (sub)models themselves, its implementation requires a new version of the foundations of the language system on which SIMTHESys is founded. This new version has been obtained by employing a minimal modification and keeping full compatibility with all the existing solvers and developed models, thanks to the metamodeling-based architecture of the language system. SIMTHESys models are described through user-defined formalisms. Formalisms are described utilizing the *SIM-THESys metaformalism*, which defines the primitives by which formalism elements may be defined in terms of *behaviors* and *properties*, as described in [19]. On the side of solving engines, SIMTHESys internals are made available using *solver helper interfaces*, which may be freely developed to hook new solving engines. The metaformalism also defines the possible types of formalism elements. Consequently, to extend SIMTHESys with rewriting features, the type system has been increased by adding the type *variable*, inspired by the homonymous type of the Pascal language. A *variant* is a structured data, accompanied by a field of enumeration type. Depending on the value contained in the enumeration, the schema of the structured data might change. *variant* are *atomic*, in the sense that, whenever the value inside the enumeration field changes, the content of the structured data becomes invalid: for this reason, a change of the enumeration value forces all the other values to be redefined. This led to version 2.0 of the SIMTHESys metaformalism. The new type is meant to accommodate different alternative sets of SIMTHESys values, which are mandatorily updated by the SIMTHESys runtime as a whole, and each of the alternative sets, which may have a different composition, is assigned to a different implementation of a (sub)model to keep its state. When parsing a rewriting-enabled model, rewriting is executed, and the alternative sets are generated as a consequence and incorporated in the runtime data structures built by the related SIMTHESys generated solver after model parsing. The SIMTHESys runtime is also extended by including rewriting features to generate and manage the switch between alternative (sub)model implementations during the solution process. The new features are accessible via the *Rewritable* solver helper interface, enabling users to implement rewriting for transformations and late binding by creating custom-solving engines. To test the solution employing a state-of-the-art reference for the new feature, a semi-incorporated prototypical implementation of rewriting has been set up using `Maude`, here used as an external solving engine manually incorporated in-the-loop.

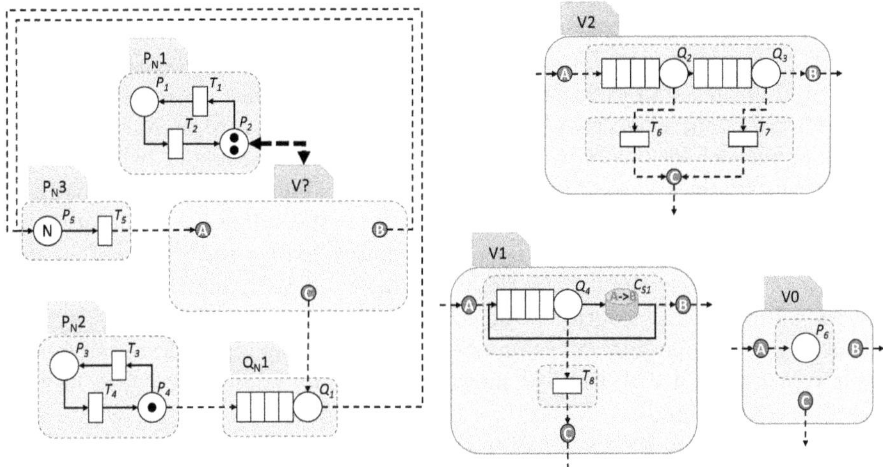

**Fig. 1.** The model for the case study system.

## 6   A Case Study

The case study concerns a server with two computing facilities, both of which are subject to faults and can be repaired, and is an evolution of the example described in [19], in which the implementation in SIMTHESys is detailed in terms of formalism definitions. Facilities have buffers for incoming requests and stop accepting requests if no buffer is available. A front-end dispatches requests to the facilities, and the second facility will also process as next jobs sent to the first, but which are not processed before a timeout. Requests are served in two stages, which, when everything works correctly, are executed by two servers working in tandem. To cope with time instants with a high arrival rate, whenever the time spent in the queue by a request becomes longer than a given threshold, it is offloaded to a third backup server. All three nodes are subject to maintenance periods in which they are not available. When one of the nodes of the tandem system is not available, the remaining one will take care of both stages. The system is modeled with three Stochastic Petri Nets (SPN) submodels (called, respectively, $P_N1$, $P_N2$ and $P_N3$), for the fault/repair process and requests injection, and two Multiclass Queuing Networks (MQN) submodels with exponentially distributed service times ($Q_N1$ and $V?$), for the three service nodes (see Fig. 1).

Multiformalism interactions between SIMTHESys submodels are implemented employing a *SIMTHESys bridge formalism* as formalism elements which can interact soundly with elements of MQN and SPN: the arc from the SPN place $P_4$ to a MQN queue $Q_1$ acts as an extension of enabling arcs in SPN; the arc from a SPN transition to a MQN queue acts by enqueuing a job when the SPN transition fires; the arcs from a MQN queue (either $Q_1$, $Q_3$ or $C_{S1}$) to the SPN place $P_5$ produce a token in the place when the queue terminates processing a job; the arcs from a MQN queue to a SPN transition (i.e. $Q_2 \rightarrow T_6$, $Q_3 \rightarrow T_7$ or $Q_4 \rightarrow T_8$) act as an extension of transition enabling in SPN, and subtract a job from the queue when they fire. The left part of the figure presents the main model: submodel $P_N3$, composed of place $P_5$ and a transition $T_5$, which defines the arrivals of the requests to the system. The main service is the rewritable model identified by $V?$, while the backup node is the submodel $Q_N1$ composed of a single queue $Q_1$. The subsystems are controlled by the Petri net models $P_N1$ and $P_N2$: while the backup node has an on-off behavior ($P_N2$ and $Q_N1$), the main subsystem is controlled by a simple Petri net with two states and two tokens ($P_N1$). Depending on the number of tokens in the place $P_2$, the submodel $V?$ is replaced by $V2$, $V1$ or $V0$ shown on the right side of the figure. If both servers are available (two tokens in $P_2$), the submodel $V2$ is used, where each server is represented by an individual queue ($Q_2$ and $Q_3$). If one node is down (one token in $P_2$), the submodel is replaced by the single queuing station $Q_4$ (submodel $V1$): the stages become two classes of customers. If a job in the first stage finishes, it enters the class switch ($C_{S1}$) and reenters the server with the class representing the second stage; otherwise, it leaves the system. If both nodes are down, no service is performed, and the system is stopped. In this case, the queue is replaced by a colored place ($P_6$ of the submodel $V0$), in which the color of the tokens represents the current stage of the corresponding job.

# 7    Using Maude as Rewriting System for SIMTHESys

In this section, we discuss the use of Maude and its runtime support as a rewriting engine for SIMTHESys. The advantages of this choice are multiple: flexibility and modularity of the modeling, efficiency, soundness, ease of integration, and upgrade. We inherit the Maude facilities for formal verification; in addition, we can exploit the quantitative analysis capability based on the association of a Markov process to Maude executable modules [8,9]. After introducing Maude, we will intuitively describe the approach to multiformalism modeling through the case study. We skip many technical details and refer for a complete description to https://github.com/lgcapra/rewpt/tree/main/multiformalism.

## 7.1    The Maude System

Maude [12] is an expressive, high-performance, purely declarative language with rewriting logic semantics [5]. The Maude runtime provides various facilities for model checking, verification of LTL formulae, and symbolic reachability. Maude has served as a logical framework for various other formalisms, including Petri nets, Automata, and Process Algebras, which, despite their strength, do not possess the essential features needed for intuitively defining adaptable systems.

Maude syntax is based on *equations* and *rules*. Each side of a rule or equation is a *term* of a certain *kind*, which may involve variables. Rules and equations operate through intuitive rewriting, where instances on the left side are replaced with instances on the right.

A *functional* module defines operations using equations as simplifications. It outlines a *equational theory* $(\Sigma, E \cup A)$ of membership equational logic : $\Sigma$ is the signature that encompasses the declaration of *sorts, subsorts, kinds*[2] and *operators*; $E$ contains equations and membership axioms; and $A$ contains the operator equational attributes (e.g., *assoc, comm, ide*). The mathematical model of $(\Sigma, E \cup A)$ is the *initial algebra* $T_{\Sigma/E\cup A}$, formed by the equivalence classes of the relation induced by $E \cup A$ in the ground-term algebra $T_\Sigma$. Under the (modulo-$A$) confluence, subsort decreasing, and termination conditions on $(\Sigma, E)$ , any ground term is rewritten in a unique canonical form that has the least sort in the sub-sort PO. The canonical term algebra is isomorphic to the initial algebra, ensuring consistency between mathematical and operational semantics.

A *system* module includes *rewrite rules* representing local transitions in a concurrent system. It defines a *rewrite theory* [5] $\mathcal{R} = (\Sigma, E \cup A, R)$. Here, $(\Sigma, E\cup A)$ acts as the equational theory below, and $R$ is a set of rewrite rules. This theory captures the behavior of a concurrent system, with $(\Sigma, E \cup A)$ defining the algebraic structure of the states and $R$ describing the concurrent transitions. The initial model of $\mathcal{R}$ provides each kind $k$ with a labeled transition system (TS) where states are elements of $T_{\Sigma/E\cup A,k}$ and state transitions occur as $[t] \overset{[\alpha]}{\to} [t']$, with $[\alpha]$ denoting an equivalence class of rewrites. The property of *coherence* [12] ensures that a strategy reducing terms to canonical forms before applying rules

---

[2] Kinds are implicit equivalence classes formed by connected components of sorts under the subsort partial order. Terms of a kind without a sort denote *errors*.

(adopted by the `Maude` rewrite engine) is sound and complete. The following case study demonstrates how these rewriting capabilities, implemented via Maude, enhance the modeling power of SIMTHESys.

## 7.2  Case Study Formalization and Analysis

The multiformalism encoding in `Maude` utilizes a modular, compact, and extensible hierarchy of modules (*parametrized*). Our elegant solution highlights advanced `Maude` module operations, facilitating the reuse of formalisms. The entire hierarchy is available at https://github.com/lgcapra/rewpt/tree/main/multiformalism. A parametrized module utilizes type parameters by *theories* defining syntactic and semantic properties for parameter modules, and *views* linking a theory to a target module or theory with a mapping of sorts and operators of the theory. To enhance readability, we provide code excerpts.

As usual, the formalization is based on multisets, which are implemented both as a commutative monoid or as compact weighted sums (module `BAG{X :: TRIV}`), for efficiency reasons. The core idea is that a multiformalism model combines diverse components (e.g., Petri nets, queue networks, BPMN, activity diagrams) to maintain a distributed state notion. This is formalized by the theory `STATE`, which requires two linked *sorts*, `State` and `LocState`, through the *subsort* relationship and an associative-commutative (AC) juxtaposition. The theory defines the parameter of the module `NETWORK{S :: STATE}`, detailing the Abstract Data Type (ADT) for a multiformalism model.

The model's abstract structure is defined using the sort `Network` and the AC juxtaposition `_,_` with `emptyNetW` as the identity. A multiformalism model is a multiset (commutative monoid) of different `Network` components, called nodes. By instantiating node types through subsort relationships, we can build a model with various components interacting via shared state elements (*places*). A term `N : S` of sort `NetSys` comprises a `Network` and a `State`, representing a network of connected nodes with a distributed *state*.

In the case study, the network nodes use a Petri net-like state (*marking*), a multiset of places represented via the `PBAG{PL :: TRIV}` module (The trivial theory `TRIV` declares the sort `Elt`). The module's type parameter indicates place labels, here indexed by naturals. The AC operator `_+_`, a constructor, describes multisets as weighted sums. For example, `3. p(1) + 1. p(2)` of sort `Pbag` from `PBAG{Nat}` represents a multiset with three places $p_1$ and one place $p_2$. The parametrized view `S-Pag{PL :: TRIV}` instantiates `STATE` as a marking. State elements correspond to elementary bags like `3 . p(1)` of sort `ElPbag < Pbag`. Using this view, we derive the ADT of heterogeneous networks for distributed state marking (module `NETWORK-MARKING{PL :: TRIV}`).

```
fth STATE is
   sorts State LocState .
   subsort LocState < State .
   op _+_ : State State -> State [assoc comm] .
endfth

fmod NETWORK{S :: STATE} is
```

```
  protecting EXT-BOOL .
  sorts Network NetSys .
  op emptyNetW : -> Network [ctor] .
  op _,_ : Network Network -> Network [ctor assoc comm prec 123 id: emptyNetW] .
  op _:_ : Network S$State -> NetSys [ctor prec 125] .
  op netw : NetSys -> Network .
  op state : NetSys -> S$State .
  vars N N' : Network . var M : S$State .
  eq netw((N : M)) = N .
  eq state((N : M)) = M .
endfm


view S-Pbag{PL :: TRIV} from STATE to PBAG{PL} is
  sort State to Pbag .
  sort LocState to ElPbag .
endv


fmod NETWORK-MARKING{PL :: TRIV} is
  protecting NETWORK{S-Pbag{PL}} .
endfm
```

*Components used in the Case-study.* Two types of nodes are used: Stochastic Petri Nets (SPN) and Multi-class Queue Networks (MQN). For the former type, we reuse the signature of (rewritable) SPN [9], which is based, in turn, on that of (rewritable) Place-Transition (PT) nets with inhibitor edges provided in [7,10]. The SPN formalization is available at https://github.com/lgcapra/rewpt.

*SPN.* The module SPN-NODE{PL :: TRIV} defines SPN nodes by importing NETWORK-MARKING and SPN-SIG, using PL as an actual parameter. Importing SPN-SIG in protecting mode retains its semantics, while NETWORK-MARKING in extending mode adds new values to Network without identifying existing ones.

SPN transitions (Tran terms) are defined by labels linked to adjacency lists as Pbag triples [I, O, H] (input, output, inhibitor). Labels consist of a descriptive tag (here a String), a rate parameter (Float) for a negative exponential delay, and a firing policy (Nat). The ground term

```
    t("a", 1.5, 0) |-> [1 . p(1) + 2 . p(2), 1 . p(1), 2 . p(1)]
```

defines a transition with label "a", an exponential rate $\mu = 1.5$, and infinite-server type. It requires exactly one token in place $p_1$ and at least two tokens in place $p_2$ to fire, removing two tokens from $p_2$. The PT net of an SPN (Net) is modularly defined using AC juxtaposition ; and subsort relation Tran < Net.

The firingRate operator helps define marking-dependent rates based on the enabling degree $(ed(t,m))$, indicating simultaneous transition occurrences in a marking. The transition exponential rate is $\mu \cdot ed(t,m)$ for the infinite server policy (0), and $\mu \cdot min(ed(t,m),k)$ for the $k$-server policy $k > 0$.

To reuse SPN transitions in multiformalism, we have to encapsulate the pre-defined SPN signature (SPN-SIG) and the network ADT (NETWORK) into a new module (SPN-NODE), incorporating the subsort Tran < Network.

Using this approach, we can add new types of nodes in a standardized way.

```
fmod SPN−NODE{PL :: TRIV} is
  extending NETWORK−MARKING{PL} .
  protecting SPN−SIG{String, PL} .
  subsort Tran < Network .
endfm
```

*Multi-class Queue Networks.* MQN are built from elementary queues (single servers), i.e., terms of sort `ElQueue` as defined in module `EL-QUEUE{PL :: TRIV}`, comprising a `Place` and an `Float` (exponential service rate), denoted by `P @ F`.

The MQN signature is in module `QUEUE{PL :: TRIV}` including `LIST` and `PBAG` in `protecting` mode. A simple MQN, a term of sort `SimpleQ`, is a non-empty list of elementary queues (term of sort `NeList{ElQueue}`) followed by a `Place`, noted as `NeL > P`. A general MQN, a term of sort `queue`, has a `SimpleQ` prefixed by two `Pbag` terms in `[]`, representing the queue input and inhibitor conditions in the context. The subsort relationship `SimpleQ < Queue` holds.

This notation allows flexible representation of any MQN. The `SimpleQ` term `p(1) @ 1.0 p(2) @ 1.5 > p(3)` describes a two-class queue, while the `Queue` term `[2. p(1), nilP] p(4) @ 2.0 > p(5)` signifies a single-class queue that needs two tokens at `p(1)`. MQN nodes connect to a heterogeneous network via the `QUEUE-NODE` module, which integrates the `NETWORK-MARKING` ADT and `QUEUE`. Network nodes combine arbitrarily (,): If q1 and q2 are `Queue` terms, and q1's endpoint aligns with q2's start, then q1, q2 represents their sequence, with time semantics distinct from an MQN merging q1 and q2.

*Node Dynamics.* The behavior of a multiformalism model is defined by system modules linked to node types, such as `SPN-NODE-SYS` and `QUEUE-NODE-SYS`. These modules use conditional rewrite rules to set the timed semantics of network nodes, locally adjusting the state of a `SysNet` term N : M, which represents interconnected nodes with a multiset of places. State changes occur due to an SPN transition or an MQN client service event that reaches a server or exits the MQN. The free variable `rate` is linked through matching equation := to an expression defining the event's time semantics, which for an MQN, depends on the client ratio at a place versus the MQN's total population. This allows for automated creation of the CTMC generator matrix according to the method [8].

```
mod SPN−NODE−SYS{PL :: TRIV} is
  including SPN−NODE{PL} .
  var N : Network . vars B B' : Pbag . var K : NzNat . var rate : Float . var T : Tran .
  crl [spn−t] : (N , T) : B => (N , T) : B' if enabled(T, B) /\ B' := firing(T, B) /\
      rate := firingRate(T, B) .
endm
```

*Case Study's Model* . The multiformalism model shown in Fig. 1 is encoded in a system module (`MQN-SPN`) that integrates the two modules expressing node dynamics and also defines the rules for structural transformations of a network component (marked $V_i$), depending on the marking of the place $p_2$. (The module is listed in the appendix.) This is a conventional method for constructing a multiformalism model with reconfigurable components.

*Experimental Evidence.* Table 1 presents some experimental results related to the `Maude` encoding of the case study. They refer to the transition system (TS) generated by the parametric term `|netsys(N)|` for increasing values of $N$ (the initial population of the network). For example, the following command – which has no solutions for any $N$– searches for final states throughout the TS. Using different shapes of the same command, we can check other base properties, e.g. preservation of the population in the system. Note that actual performance measurement (i.e., system response time) has not been reported since it depends only on the parameters of the model, and their computation time is independent of the actual values provided. Instead, the table shows the relative scalability of the approach, emphasizing that middle-to-large setups can be analyzed in a few dozens of minutes on a PC with an 11th-Gen Core i5 and 32 GB RAM.

```
Maude> search in MQN-SPN : netsys(N) =>! F:NetSys .
```

**Table 1.** Transition System build of the case-study

| $N$ | # states | build time (sec) |
|---|---|---|
| 10 | 5.148 | 4 |
| 20 | 31.878 | 30 |
| 30 | 98.208 | 53 |
| 40 | 222.138 | 130 |
| 50 | 421.668 | 290 |
| 60 | 714.798 | 413 |
| 70 | 1.119.528 | 718 |
| 80 | 1.653.858 | 1080 |
| 90 | 2.335.778 | 2.429 |
| 100 | 3.183.318 | 3.290 |

## 8   Conclusions

This work has demonstrated how the integration of a rewrite engine into SIM-THESys significantly improves multiformalism modeling capabilities. By extending the metaformalism with a variant type and incorporating `Maude` as a rewriting system, this approach enables the dynamic transformation and late binding of submodels during the solution process. The case study discussed illustrates the practical benefits of this approach, showing improved modeling flexibility, modularity, and computational efficiency. The preliminary results confirm that the ability to dynamically reconfigure models during evaluation can mitigate issues such as state-space explosion and adapt to system dynamics more effectively. In the SIMTHESys multiformalism modeling framework, the role of `Maude` is that of a solution engine, which acts on the structure of the model at solution time and supports the dynamic reconfiguration of the model representation at solution time, thus introducing new features without the need for a redesign of the conceptual framework nor of the SIMTHESys*ER* solvers generation tool. Future work will aim at further optimization for a smooth integration of `Maude` in the framework, the exploration of automated strategies for model transformations, aiming to improve SIMTHESys in modeling and analyzing complex heterogeneous systems, and testing and evaluating further methodological development which may result from the integration of rewriting in SIMTHESys. We aim to enhance the `Maude` multiformalism framework by integrating compositional operators to emphasize network symmetries, intending to derive a lumped Markov process, akin to the methodology in [10] for rewritable SPN.

## Appendix

The following listing shows the `Maude` encoding of the case study and has been included for the convenience of the readers.

```
mod MQN-SPN is
  inc SPN-NODE-SYS{Nat} .
  inc QUEUE-NODE-SYS{Nat} .
  var K : NzNat . *** model parameter
  vars N N' N'' : Network . var S : Pbag .
  ops t0 t1 t2 t3 t4 t5 t6 : -> Tran . *** aliases
  ops eq1 eq2 eq3 : -> ElQueue . *** aliases
  eq eq1 = p(7) @ 1.0 . *** conditioned single-class queue
  eq eq2 = p(1) @ 1.5 .
  eq eq3 = p(6) @ 2.5 .
  ops q1 q2 q3 q23 : -> Queue . *** aliases
  eq q1 = [1 . p(5), nilP] eq1 > p(0) .
  eq q2 = [2 . p(2), nilP] eq2 > p(6) .
  eq q3 = [2 . p(2), nilP] eq3 > p(0) .
  eq q23 = [1 . p(2), nilP] q1(q2) q1(q3) > out(q3) . *** MQN alias
  op network : -> Network . *** alias
  op netsys : NzNat -> NetSys . *** alias
  eq t0 = t("start", 1.0, 1 ) |-> [1 . p(0), 1 . p(1), nilP] .
  eq t1 = t("switch1", 0.5, 1 ) |-> [1 . p(3), 1 . p(2), nilP] .
  eq t2 = t("switch2", 0.05, 1 ) |-> [1 . p(2), 1 . p(3), nilP] .
  eq t3 = t("on", 2.0, 1 ) |-> [1 . p(4), 1 . p(5), nilP] .
  eq t4 - t("off", 1.0, 1 ) |-> [1 . p(5), 1 . p(4), nilP] .
  eq t5 = t("rem1", 1.0, 1 ) |-> [1 . p(1), 1 . p(7), nilP] .
  eq t6 = t("rem6", 1.5, 1 ) |-> [1 . p(6), 1 . p(7), nilP] .
  op V : NzNat -> [Network] [memo] . *** variable component (depends on p(2))
```

eq $V(2)$ = q2 , q3 , t5 , t6 . *** sequential composition: "out" of q2 is "in" for q3
eq $V(1)$ = q23 , t6 . *** hybrid component (MQN + tran)
eq $V(0)$ = p(eq2) @ 0.0 p(eq3) @ 0.0 > out(q3) . *** "dead" queue
eq network = t0 , t1 , t2 , t3 , t4 , t5 , t6 , q1 , $V(2)$ .
eq netsys(K) = network : K . $p(0)$ + 2 . $p(2)$ + 1 . $p(5)$ .
*** network rewriting
crl [V2>V1] : N : S => N' , $V(1)$ : S if S[$p(2)$] = 1 /\ N'' , N' := N /\ N'' = $V(2)$ .
crl [V2>V0] : N : S => N' , $V(0)$ : S if S[$p(2)$] = 0 /\ N'' , N' := N /\ N'' = $V(2)$ .
crl [V1>V2] : N : S => N' , $V(2)$ : S if S[$p(2)$] = 2 /\ N'' , N' := N /\ N'' = $V(1)$ .
crl [V1>V0] : N : S => N' , $V(0)$ : S if S[$p(2)$] = 0 /\ N'' , N' := N /\ N'' = $V(1)$ .
crl [V0>V2] : N : S => N' , $V(2)$ : S if S[$p(2)$] = 2 /\ N'' , N' := N /\ N'' = $V(0)$ .
endm


# References


1. Ardagna, D., et al.: Predicting the performance of big data applications on the cloud: D. ardagna et al. J. Supercomput. **77**(2), 1321–1353 (2021)
2. Barbierato, E., Gribaudo, M., Iacono, M.: Modeling hybrid systems in SIMTHE-Sys. Electron. Notes Theor. Comput. Sci. **327**, 5–25 (2016)
3. Barbierato, E., Gribaudo, M., Iacono, M., Jakóbik, A.: Exploiting CloudSim in a multiformalism modeling approach for cloud based systems. Simul. Model. Pract. Theory **93**, 133–147 (2019)
4. Bause, F., Buchholz, P., Kemper, P.: A toolbox for functional and quantitative analysis of DEDS. In: Puigjaner, R., Savino, N.N., Serra, B. (eds.) TOOLS 1998. LNCS, vol. 1469, pp. 356–359. Springer, Heidelberg (1998). https://doi.org/10.1007/3-540-68061-6_32
5. Bruni, R., Meseguer, J.: Generalized rewrite theories. In: Baeten, J.C.M., Lenstra, J.K., Parrow, J., Woeginger, G.J. (eds.) ICALP 2003. LNCS, vol. 2719, pp. 252–266. Springer, Heidelberg (2003). https://doi.org/10.1007/3-540-45061-0_22
6. Bucci, G., Sassoli, L., Vicario, E.: Oris: a tool for state-space analysis of real-time preemptive systems, pp. 70–79 (2004)
7. Capra, L.: Rewriting logic and petri nets: a natural model for reconfigurable distributed systems. In: Bapi, R., Kulkarni, S., Mohalik, S., Peri, S. (eds.) ICDCIT 2022. LNCS, vol. 13145, pp. 140–156. Springer, Cham (2022). https://doi.org/10.1007/978-3-030-94876-4_9
8. Capra, L.: Associating a Markov process with Maude executable modules. In: Proceedings of the 15th International Conference on Simulation and Modeling Methodologies, Technologies and Applications, pp. 106–116. SciTePress (2025)
9. Capra, L., Gribaudo, M.: A lumped CTMC for modular rewritable PN. In: Doncel, J., Remke, A., Di Pompeo, D. (eds.) Computer Performance Engineering. EPEW 2024. LNCS, vol. 15454, pp. 106–120. Springer, Cham (2025). https://doi.org/10.1007/978-3-031-80932-3_8
10. Capra, L., Köhler-Bußmeier, M.: Modular rewritable Petri nets: an efficient model for dynamic distributed systems. Theor. Comput. Sci. **990**, 114397 (2024)
11. Ciardo, G., Miner, A.S.: SMART: the stochastic model checking analyzer for reliability and timing. In: International Conference on the Quantitative Evaluation of Systems, pp. 338–339. IEEE, Los Alamitos, CA, USA (2004)

12. Clavel, M., et al.: All About Maude - A High-Performance Logical Framework. LNCS, vol. 4350. Springer, Heidelberg (2007). https://doi.org/10.1007/978-3-540-71999-1

13. Deavours, D.D., et al.: The Möbius framework and its implementation (2002)

14. Franceschinis, G., Gribaudo, M., Iacono, M., Marrone, S., Moscato, F., Vittorini, V.: Interfaces and binding in component based development of formal models. In: Proceedings of the 4th International ICST Conference on Performance Evaluation Methodologies and Tools, pp. 44:1–44:10. VALUETOOLS '09, ICST, ICST, Brussels, Belgium, Belgium (2009)

15. Franceschinis, G., Gribaudo, M., Iacono, M., Mazzocca, N., Vittorini, V.: DrawNET++: model objects to support performance analysis and simulation of systems. In: Field, T., Harrison, P.G., Bradley, J., Harder, U. (eds.) TOOLS 2002. LNCS, vol. 2324, pp. 233–238. Springer, Heidelberg (2002). https://doi.org/10.1007/3-540-46029-2_18

16. Gianniti, E., Rizzi, A.M., Barbierato, E., Gribaudo, M., Ardagna, D.: Fluid petri nets for the performance evaluation of mapreduce and spark applications. ACM SIGMETRICS Perform. Eval. Rev. **44**(4), 23–36 (2017)

17. Gribaudo, M., Iacono, M.: Theory and application of multi-formalism modeling (2013). https://doi.org/10.4018/978-1-4666-4659-9

18. Iacono, M., Barbierato, E., Gribaudo, M.: The SIMTHESys multiformalism modeling framework. Comput. Math. Appl. **64**, 3828–3839 (2012)

19. Iacono, M., Gribaudo, M.: Element based semantics in multi formalism performance models. In: Proceedings of the 18th IEEE/ACM International Symposium on Modeling, Analysis and Simulation of Computer and Telecommunication Systems, MASCOTS 2010, pp. 413–416 (2010)

20. Lara, J., Vangheluwe, H.: AToM$^3$: a Tool for Multi-formalism and Meta-modelling. In: Kutsche, R.-D., Weber, H. (eds.) FASE 2002. LNCS, vol. 2306, pp. 174–188. Springer, Heidelberg (2002). https://doi.org/10.1007/3-540-45923-5_12

21. Moscato, F., Flammini, F., Lorenzo, G.D., Vittorini, V., Marrone, S., Iacono, M.: The software architecture of the OsMoSys multisolution framework. In: ValueTools '07: Proceedings of the 2nd International Conference on Performance Evaluation Methodologies and Tools, pp. 1–10 (2007)

22. Raiteri, D.C., Iacono, M., Franceschinis, G., Vittorini, V.: Repairable fault tree for the automatic evaluation of repair policies. In: DSN, pp. 659–668. IEEE (2004)

23. Sanders, W.: Integrated frameworks for multi-level and multi-formalism modeling. In: Proceedings 8th International Workshop on Petri Nets and Performance Models, pp. 2–9 (1999)

24. Trivedi, K.S.: Sharpe 2002: symbolic hierarchical automated reliability and performance evaluator. In: DSN '02: Proceedings of the 2002 International Conference on Dependable Systems and Networks, p. 544. IEEE, Washington, DC, USA (2002)

# Risk Assessment Models in Hydroelectric Plants

Davide Cerotti[1(✉)] [iD], Lavinia Egidi[1] [iD], Daniele Codetta-Raiteri[1] [iD],
Marzio Alfio Pennisi[1] [iD], Giuliana Franceschinis[1] [iD], Andrea Bobbio[1] [iD],
Luigi Portinale[1] [iD], Marco Cappicciola[2], Edoardo Martino L'Aurora[2],
and Noemi Silicato[2]

[1] DISIT, Università del Piemonte Orientale, Alessandria, Italy
`{davide.cerotti,lavinia.egidi,daniele.codetta,marzio.pennisi,`
`giuliana.franceschinis,andrea.bobbio,luigi.portinale}@uniupo.it`
[2] Enel Green Power, Rome, Italy
`{marco.cappicciola,edoardomartino.laurora,noemi.silicato}@enel.com`

**Abstract.** Recently, renewable energy including hydroelectric power, is
gaining relevance. Hydropower plant operation is influenced by many factors, so risk assessment becomes a key process to schedule maintenance
and guarantee the production of the necessary amount of energy. We
define a methodology for risk analysis considering failure processes and
repair actions, and based on the definition of component and subsystem
hierarchy, interdependencies, failure and repair parameters. Hierarchy
higher levels are modelled by Fault Trees, while lower levels are modelled by a dependency graph. Component Health Status (CHS©) is a
wear index evaluated by experts, from which failure and repair distributions are derived through a fitting procedure. Producibility levels and
risk matrix are the output of our methodology, which is applied, as a
case study, to a simplified hydropower plant.

**Keywords:** hydropower plant · risk analysis · failure · repair ·
maintenance · fault tree · dependency graph · component health status

## 1 Introduction

Energy generation plants are among the most critical systems in our society
because of the hunger for energy that characterizes our way of life. In particular,
with climatic change and a need for an economy that moves away from fossil
fuels, renewable energy such as hydroelectric power is becoming more and more
central.

The continued operation of hydropower plants is not an easy task because of
the complexity of the systems and because of the many factors that come into
play. So risk assessment becomes fundamental as a basis for scheduling necessary maintenance and understanding producibility fluctuations for a cooperative
interplay of different plants to ensure the required energy production.


L. Carnevali and J. Doncel (Eds.): EPEW 2025, LNCS 15657, pp. 17–28, 2026.
https://doi.org/10.1007/978-3-032-16345-5_2

In this context, the line of research presented in this paper is a cooperation between Università del Piemonte Orientale (UPO) and Enel Green Power (EGP), an Italian company operating throughout the world with their renewable energy plants and working towards electrification and sustainable energy.

This collaborative work originates from the necessity to formalize a method with a solid theoretical foundation that, prospectively, will enable EGP to conduct risk analyses considering failure and repair processes (with a potentially very fine-grained level of detail on the plant's constituent components) as well as extraordinary maintenance plans. The starting point is a substantial effort involving the hierarchical cataloging of the involved components, the definition of interdependencies, and the identification of parameters characterizing failure processes (accounting for deterioration over time) and repair processes. The ultimate objective is to be able to evaluate risk with respect to various indicators, although currently only the impact related to "producibility" has been considered.

Our contribution is a hierarchical approach aiming at breaking down the high complexity of the plants. Systems, subsystems and single pieces of equipment are catalogued in a hierarchical fashion, level 1 of the hierarchy being the whole plant and the highest level the single component. We take advantage of this hierarchy by identifying a specific level above which the subsystem is modelled by Fault Trees, whereas dependencies among subsystems of lower level are modelled with a dependency graph.

The wear condition of each component and, by extension, of any subsystem is expressed in terms of the Components Health Status ($CHS^{©}$). Currently the parameters used in the models are determined by expert evaluation within EGP. But medium- to long-term work aims at integrating a "pipeline" that allows for the automatic derivation of these parameters from historical data series across the vast array of plants managed by EGP.

The methodology we describe in this paper has been applied by EGP for risk assessment in all of their plants throughout the world. The EGP hydroelectric fleet consists of more than 700 operating hydropower plants distributed in 14 European and Latin American different countries, for a total of more 27 GW installed capacity.

The paper is organized as follows. Section 2 discusses related research and differences with respect to our work. Section 3 presents the case study and its models. Section 4 explains fitting of failure and recovery distributions. Section 5 shows the computation of producibility levels and risk matrix. Finally, Sect. 6 concludes the paper, discussing results and future work.

## 2     Related Work

The literature related to risk assessment and availability of hydroelectric power plants is very extensive and varied, given the criticality of the context. We report here some works to frame our contribution in the wider research context.

The paper [1] proposes a Monte Carlo simulation based on real data from a hydropower plant in Brasil to obtain availability projections. This is aimed at optimizing maintenance also in view of regulatory aspects and studying the impact of shutdowns due to mandatory maintenance. The paper uses historical data of forced shutdowns. The context of our work is different in that our availability predictions take their moves from knowledge of the state of the system and its components.

The paper [2] is interesting in our context since it starts from a representation of functional dependencies among the different systems, subsystems and components of the plant. From the initial availability information, a Markovian propagation method determines availability in time, having obtained reliability data of each component from the specific organization. In the conclusions, they mention as future work considering degraded initial states of the basic components (which in our work is captured by the CHS©). But to the best of our knowledge, this hasn't been pursued.

Simulation has been used in several works. For instance, in [3] an analysis to expose faults in hydropower plants has been conducted with the use of simulations on MATLAB/Simulink. The study focuses on the detection of two specific faults and on the performance of the plant in challenging conditions. Our point of view is different since we are not aiming at detecting specific faults but rather at evaluating the availability and the producibility under normal operating conditions. In [4], the authors use simulation to address fault tolerance, with the restriction to specific components. These studies go into deeper detail but don't give us the generality and broad view that we need.

The work [5] relates to our work in that its aim is determining availability over time. It uses a method based on binary trees to evaluate the distance of one-year operating data from the time-series anomaly metrics. Using this model, the authors address maintenance of small hydropower systems. We need a method that scales well, to address the huge variety of EGP plants.

The paper [6] proposes a system to monitor in real time faults of a hydropower plant. The system is composed of a monitoring system to acquire real time data and a support vector machine classifier for the diagnosis. The approach has raised much interest and there has been lively activity in the research field. Of course, the necessity of sensors for real time analysis changes the setting with respect to our context. The results presented in [7] can be seen as a recent development of this line of research. It focuses on the risk assessment of hydroelectric generating units and it proposes the use of fuzzy classifiers to assess risk based on reference benchmark models and data from sensors. Again, the assessment is based on data from sensors, and in this case only the generating units are under scrutiny.

In [8], the objective is determining investment risk related to mini hydropower plant projects. We mention it here because the proposed methodology combines Fault Tree Analysis (FTA) and risk matrices. There is actually a rich literature on risk assessment related to hydroelectric power plant projects, addressing investment, environmental, social, and anthropological risk factors and impacts. Papers that take this broader point of view are surveyed in [9]. Other papers look at risk and impact on productivity in the case of extreme events, e.g. [10].

In [11] the credit risk of financing a small hydropower plant (SHPP) in Serbia is assessed via the failure mode and effects analysis (FMEA). Considering the epistemic uncertainty [12] FMEA-DST methodology was applied to identify the risk events requiring mitigation strategies to be employed by credit risk managers.

## 3  Hydroelectric Plant Model

The purpose of this work was the risk assessment of hydroelectric power plants in terms of the expected reduction in energy produced during a given period. The plant consists of subsystems dedicated to performing various functions, such as collecting water in basins, channeling it toward turbines through penstocks, etc., where each subsystem is, in turn, composed of different components. Each component can fail, causing a complete loss of its functionality, and subsequently be repaired, returning to a fully operational state. Due to the difficulty in obtaining detailed and exhaustive information on the functional dependencies among all the various components of the plant, in agreement with EGP, it was conservatively assumed that the failure of any single component causes the failure of the entire subsystem to which it belongs. Instead, the functional relationships between subsystems are more complex. For example from a production standpoint, a dam failure causes a loss of the water collected in the reservoir, preventing it from being channeled to the turbines and halting energy production, even if the turbines are still operational. Thus, we can identify non productive elements that can affect the energy production.

In this work the description of a given plant, including its subsystems, components, their characteristics and functional relationships are acquired automatically by a data platform from a company database. The status of each component is monitored and its $CHS^{\copyright}$ is periodically collected and saved in the database. From the previous description, we can recognize a hierarchical structure that can be easily generalized and exploited. This section outlines the proposed hierarchical model and presents an example derived from a simplified plant used as a case study.

### 3.1  Hierarchical Model

The description of a given plant encompasses various levels (e.g. plant, subsystem, components), but those of interest are at *Subsystem Level* and those below it. Among the subsystems, some are the actual energy generation units for which the maximum producibility is known. The plant has a potential producibility that is a function of the producibility of the generation units at the Subsystem Level (in the simplest case, the sum of the producibilities of the generation units). The availability of a given subsystem depends on two factors (in an OR relationship): the state of its components and the state of other subsystems.

The time-to-failure parameters depend on the level of deterioration of the elements. Currently, it is expressed through the $CHS^{\copyright}$ defined by experts at the component level.

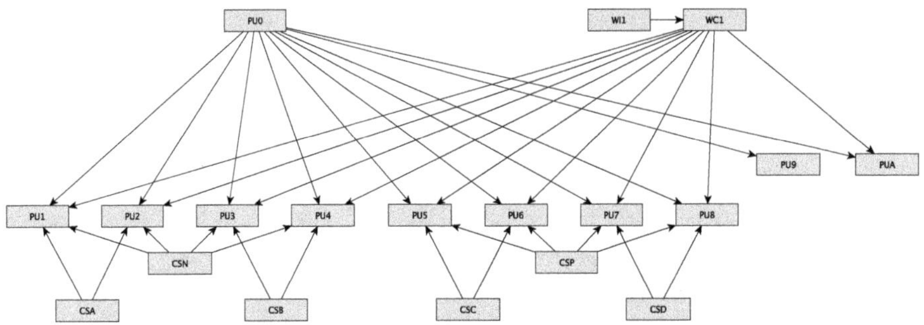

**Fig. 1.** Dependency graph between subsystems.

A challenge to the construction of the model is posed by the dimensions or complexity of the system, which depends on both the number of components and functional blocks and on their correlations and reciprocal dependencies. Real plants may have hundreds of components and dependencies between them, thus we take a hierarchical approach: dependencies between subsystems will be represented with a dependency graph, whereas dependencies between components of the same subsystem will be modeled by Fault Trees. In this framework we assume that a failure in a component of a subsystem does not affect the condition of a component in other subsystems.

## 3.2   Dependency Graph

The dependency graph represents the relationships between subsystems through directed arcs of the "depend-on" type. The direction of the arrows connecting the various blocks indicates a dependency relationship regarding the failure of the involved blocks. In particular, the failure of the source node causes the destination node to fail. Note that we consider only acyclic dependancy graphs, and therefore the dependency chain is always of limited length.

In the dependency graph of Fig. 1, nodes $PU1$ through $PU8$ are the actual production units; node $WI1$ represents the water intake (the water reservoir, including the dam, water intakes and gates), node $WC1$ the water conveyance works (canals, penstocks, etc.). Nodes $CSX$ (with $X=A$, $B$, $C$, $D$, $N$, $P$) are common service units that affect various groups of production units as indicated in the graph of Fig. 1. A failure of $WC1$, that causes, e.g., insufficient water flow, determines the failure of all production units $PU1$ through $PU8$. A failure in $WI1$ (lack of water, damaged dam, blocked intakes) renders $WC1$ ineffective and as a consequence, again, production is stopped at all units. Failure of $CSA$ causes the loss of production of units $PU1$ and $PU2$, failure of $CSN$ the loss of production of units $PU1$, $PU2$, $PU3$ and $PU4$, and so on. This kind of representation is a graphical transposition of dependency tables provided by EGP. In case of several arcs pointing towards the same node they are combined according to an OR logic. In this case study the evaluation of the impact of

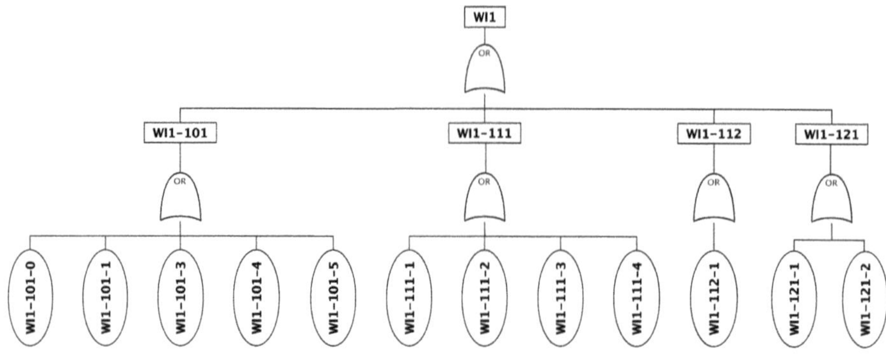

**Fig. 2.** Fault Tree describing components of subsystem $WI1$ (water intake).

failures is measured in terms of loss of "producibility" due to unavailability of one or more production unit ($PU_i$). The overall loss of producibility is thus a function of the simultaneous (un)availability of the eight production units: as a consequence there isn't a single Top Event with only two possible states (*true* or *false* with associated probabilities) but rather several degradation levels, which are function of the states of all production units.

### 3.3   Fault Tree Model

The *Fault Tree* model [13] shows how several combinations of component failures lead to the whole system failure. The model is composed of events and logic ports. Events are used to represent failures of components, or whole systems. We can consider an event as a Boolean variable: it may be false (working) or true (failed).

In the proposed model *Basic events* (oval nodes in Fig. 2) represent the component states. For example, in Fig. 2 we show a Fault Tree for the water intake, where the basic events WI1-101-0, WI1-101-1, WI1-101-3, WI1-101-4, WI1-101-5, etc. describe states of, e.g., gates, intakes, surface reservoir outlets, etc. A basic event may change its Boolean value (false or true) after a random time which can be ruled by several types of probability distribution (Sect. 4) whose typical parameters are: failure rate $\lambda$ (the inverse of mean time to failure - MTTF) and repair rate $\mu$ (the inverse of mean time to repair - MTTR). Basic events are assumed to be probabilistically independent.

*Intermediate events* (rectangle nodes in Fig. 2) represent the macro component states. For example, in Fig. 2 we have the intermediate events WI1-101, WI1-111, WI1-102, WI1-121, which represent failures of, e.g., water intake, withdrawal and conveying structures, reservoir outlet structures, etc. Finally, the *top event* (event WI1 in Fig. 2) represents the whole subsystem state. The Boolean value (false or true) of an intermediate or top event is the output of a logic port.

Logic ports are connected, by means of arcs, to several input events and one output event. The output event value depends on the input event values and the Boolean (logic) function represented by the port type: AND, OR, etc. A change of the input values determines the immediate update of the output value. In Fig. 2, only the OR port is used. For example, the basic events WI1-101-0, WI1-101-1, WI1-101-3, WI1-101-4, WI1-101-5 are the inputs of the OR port having the intermediate event WI1-101 as output. These basic events represent failures of gates and intakes, among others, that are part of the water intake (WI1-101). In turn, WI1-101 is the input, together with WI1-111, WI1-102, WI1-121, of the OR port having the top event WI1 as output.

Through specific analysis procedures, we can compute several measures on this model, including the system unavailability, which in this context, is the probability that the subsystem has failed (the top event is **true**) at a certain time.

## 3.4 Unavailability Computation

The unavailability $D(t)$ can be defined at a certain instant of time $t$, assuming that at time $t = 0$ the object is working. As $t \rightarrow \infty$, the unavailability tends to a constant value $D = \lim_{t->\infty} D(t)$, called asymptotic or steady-state unavailability. Moreover, an expected interval unavailability can be defined as the expected value of the unavailability over a given time interval.

When both the failure and recovery (repair) processes of the basic events follow exponential distributions, we have closed-form expressions to derive the unavailability. Table 1 shows the formulas for calculating the instantaneous unavailability $(D(t))$, in steady state $(D(t), t \rightarrow \infty)$, and the expected interval availability $(D_I(t))$. All such formulas can be derived from the solution of a two-state Continuous Time Markov Chain (see [13]). To compute the risk during a given interval of time (see Sect. 5) and derive the unavailability of the basic events in the Fault Tree analysis, we will use the formula in column 2, to compute the steady-state unavailability $D$, and that in column 3 for the expected interval unavailability $D_I(t)$. The former can be adopted when the analysis is conducted over a long enough period to assume that a steady state has been reached; the latter otherwise.

**Table 1.** Formulas for calculating unavailability in its different formulations. Parameters: $\lambda$ = failure rate, $\mu$ = repair rate. $MTTF = 1/\lambda$, $MTTR = 1/\mu$.

| $D(t) = 1 - A(t)$ | $D = 1 - A;\ (D(t), t \rightarrow \infty)$ | $D_I(t) = 1 - A_I(t) = \dfrac{1}{t} \displaystyle\int_0^t D(u)\, du$ |
|---|---|---|
| $\dfrac{\lambda}{\lambda + \mu} - \dfrac{\lambda}{\lambda + \mu} e^{-(\lambda+\mu)t}$ | $\dfrac{\lambda}{\lambda + \mu}$ | $\dfrac{\lambda}{\lambda + \mu} - \dfrac{\lambda}{t(\lambda + \mu)^2}(1 - e^{-(\lambda+\mu)t})$ |

Given the unavailability of the leaf nodes, it is necessary to calculate the unavailability of the upper-level nodes of the fault tree. The general formula for OR gate is given by:

$$D(t)_{OR} = 1 - \prod_{i=1}^{n} (1 - D_i(t)) \tag{1}$$

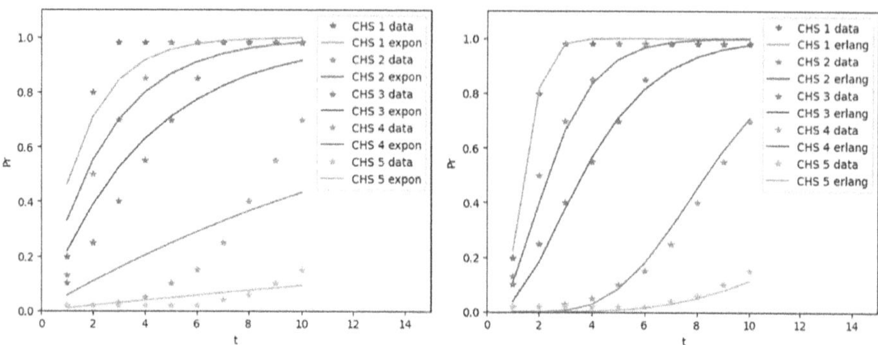

**Fig. 3.** Fitting CHS©curves using exponential (left) and Erlang distributions (right).

where $Di(t)$ is the unavailability of the input node $i$ of the OR gate and $n$ is the number of input nodes. Notice that the same formula holds for all the formulations of unavailability in Table 1.

## 4      Fitting of the Failure and Recovery Distributions

The $CHS^©$ is an integer value in the interval $[1...5]$ associated with each basic event of the Fault Tree: it expresses the component wear status and therefore is used to estimate the time to failure distribution of the component. For each $CHS^©$ different probability distributions of failure over the subsequent 10 years are considered. The increase in the probability of failure over the years depends on the specific $CHS^©$ value assigned to the component: the value 5 corresponds to the minimum increase in the probability of failure over the years, while the value 1 corresponds to the maximum increase.

We can approximate such behaviors using well-known probability distributions and derive their parameters by applying a fitting procedure that minimizes the *Mean Squared Error (MSE)*. For this work, we have assumed that the failure time of all components follows either an exponential or a $k$-stage Erlang distribution with rate $\lambda$: the fitting results are shown in Fig. 3. It can be observed that using exponential distributions yields less accurate fittings compared to the points provided by the experts, particularly for $CHS^©$ 3 and 4. Erlang distributions significantly improve the accuracy of the fitting at the cost of a larger number of stages, as shown in Table 2.

An analogous approach can be adopted to estimate the time to repair distributions of the components. In such a case, we assume that the time to repair is exponentially distributed and experts provide us an estimate of the 90% quantiles from which we can derive the distribution parameters. When both failure and repair times are exponentials we can adopt solution methods with a lower computational cost since we have closed-form expressions to derive the availability [13].

Note that the fitting only needs to be performed in the failure and recovery (repair) processes definition to obtain the parameter values that best approximate the stochastic behavior and can then be used to derive the component unavailability of basic events in the Fault Tree.

**Table 2.** Parameters of the failure distributions obtained from the fitting.

| $CHS^{©}$ | Exponential | | Erlang | | |
|---|---|---|---|---|---|
| | $\lambda$ | MSE | $k$ | $\lambda$ | MSE |
| 1 | 0.619582 | 0.100290 | 6 | 4.064847 | 0.019495 |
| 2 | 0.401259 | 0.080983 | 3 | 1.140914 | 0.040597 |
| 3 | 0.247234 | 0.105735 | 3 | 0.734382 | 0.046399 |
| 4 | 0.056711 | 0.137910 | 9 | 1.038749 | 0.032434 |
| 5 | 0.009583 | 0.026536 | 6 | 0.325151 | 0.019958 |

## 5 Producibility and Risk Matrix Computation

The impact $Im(PU_i)$ due to a failure of the subsystem $PU_i$ is equal to the loss of all of its producibility. The risk associated with the out of service of the subsystem $PU_i$, denoted as $Rk(PU_i)$, is given by the reduction in producibility due to the subsystem unavailability, therefore:

$$Rk(PU_i) = D(PU_i) \cdot Im(PU_i) \tag{2}$$

where $D(PU_i)$ is the unavailability of component $PU_i$.

Through the Fault Tree, we can calculate the unavailability for all the productive groups, $D(PU_i)$ with $i = 1, 2, \ldots 8$ in the Test Plant. Therefore, for each production group, we will have an impact equal to its producibility loss $Im(PU_i)$ and a risk $Rk(PU_i)$ given by Eq. 2.

To calculate the impact of failure of the entire plant, i.e. the reduction in producibility due to the out of service of its production groups, we need to know what the possible configurations (operational and failed) of the groups are, calculate the unavailability and the corresponding impact for each possible configuration, and then take the weighted average.

In Fig. 4 we show an example, in which we consider all the possible failure configurations in a plant with only two generation units $PU_1$ and $PU_2$. The dependencies from other subsystems are as depicted in Fig. 1. The first six columns consider the operational (*up*) or failure (*down*) state of each subsystem. Column 7 reports the impact for each configuration, and in the last column we show the formula that gives the probability of each configuration from the probabilities of availability $A(\cdot)$ and unavailability $D(\cdot)$ of the subsystems. In the last line we collect all the configurations corresponding to having a failure on any (possibly more than one) of the subsystems from which both production units depend: both $PU_1$ and $PU_2$ will stop production, independently of their own

| $PU_0$ | $WC_1$ | $CSA$ | $CSN$ | $PU_1$ | $PU_2$ | Im | Probability |
|---|---|---|---|---|---|---|---|
| up | up | up | up | down | down | 76 | $A(PU_0)A(WC_1)A(CSA)A(CSN)D(PU_1)D(PU_2)$ |
| up | up | up | up | down | up | 38 | $A(PU_0)A(WC_1)A(CSA)A(CSN)D(PU_1)A(PU_2)$ |
| up | up | up | up | up | down | 38 | $A(PU_0)A(WC_1)A(CSA)A(CSN)A(PU_1)D(PU_2)$ |
| up | up | up | up | up | up | 0 | $A(PU_0)A(WC_1)A(CSA)A(CSN)A(PU_1)A(PU_2)$ |
| $PU_0$ or $WC_1$ or $CSA$ or $CSN$ down | | | | * | * | 76 | $1-A(PU_0)A(WC_1)A(CSA)A(CSN)$ |

**Fig. 4.** Possible configurations of the case of study limited to groups $PU_1$ and $PU_2$. The symbol $*$ in the last line indicates any value (*up* or *down*) for the two subsystems $PU_1$ and $PU_2$; the last line groups $60 = (2^4 - 1) \cdot 2^2$ configurations. The impact is computed assuming that both $PU_1$ and $PU_2$ have producibility 38.

specific operational state. The impacts are computed assuming that both $PU_1$ and $PU_2$ have producibility 38.

Having calculated the unavailability and impact of the entire plant, we can calculate the risk and construct the risk matrix that visualizes the likelihood that the risk event will occur and the potential impact of such an event.

The average impact of the whole plant is obtained as the sum of the probabilities of each configuration, multiplied by its impact, divided by the probability that there is at least one unavailable productive unit. Let $\mathcal{C}$ be the set of all configurations $C$, and $C_A$ the configuration in which all production units are available; let $Pr(C)$ be the probability of configuration $C$ and $I(C)$ its impact, then the impact $(\overline{I})$ of the whole plant is given by:

$$\overline{I} = \frac{\sum_{C \in \mathcal{C}} I(C) * Pr(C)}{1 - Pr(C_A)}.$$

In the simplified example of Fig. 4, the average impact of the whole plant is obtained as follows:

$$\overline{I} = \frac{76 * Pr(PU_1 \ and \ PU_2 \ unavailable) + 38 * Pr(only \ one \ PU_i \ unavailable)}{1 - Pr(C_A)}.$$

In this formula, the probability of unavailability of both production units in the first addend is the sum of the probabilities in the first line and the last line of the table in Fig. 4; and the probability that only one production unit is unavailable $Pr(only \ one \ U_i \ unavailable)$ is the sum of the probabilities in lines 2 and 3 of the table. We omitted line 4 since the impact is 0.

We can visualize the results in a risk matrix: Fig. 5 shows the risk matrices for the example with two production units. The impact is the potential production loss depicted as a horizontal axis; the vertical axis is used to measure the probability of risk occurrence, thus the unavailability of the components or the whole plant. The risk is plotted as a dot. On the left side of Fig. 5 the risks of the plant, identified by the label $ITH$, and of its main subsystems are shown; on the right side we show only the risks of the two production units and their components. The sets of points with the same product of probability and impact are the *isorisk curves* and can be used as the boundaries of different risk categories.

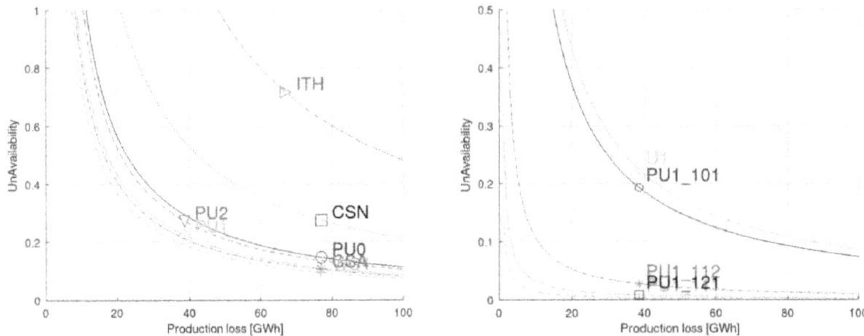



**Fig. 5.** Risk Matrices for the Test Plant.

The isorisk curves and risk matrices are especially useful to prioritize plant risks and to make decisions, such as performing repair actions on specific components or developing maintenance plans.

## 6    Conclusions and Future Work

We proposed a hierarchical method to assess risk in hydroelectric plants. Dependencies between subsystems are modelled by a dependency graph, whereas the state of their sub-components is modeled by Fault Trees. This allows us to conduct risk analyses taking into account failure and repair processes.

EGP has validated the proposed methodology, applying it to a significant set of hydropower plants; the results obtained match the expectations of experts. Our cooperation continues, and the next step is the definition of models based on simulation techniques to enable the evaluation of risk over time, taking into account the aging of components.

## 7   Copyright



## References


1. Barros de Oliveira, M.T., de Sousa Oliveira Silva, P., Oliveira, E., Marques Marcato, A.L., Junqueira, G.S.: Availability projections of hydroelectric power plants through Monte Carlo simulation. Energies **14**(24), 8398 (2021)
2. Sanchez Dominguez, J., Frias Suarez, D.G., Dominguez, D.S., Torres Valle, A., Perdomo Ojeda, M.: A comparative study of two novel reliability/availability calculation methods applied to a hydroelectric power plant. In: International Youth Conference on Energy, pp. 1–7. IEEE (2013)

3. Zabihi, A.: Assessment of faults in the performance of hydropower plants within power systems. Energy **7**(2) (2024)
4. Simani, S., Alvisi, S., Venturini, M.: Fault tolerant control of a simulated hydro-electric system. Control. Eng. Pract. **51**, 13–25 (2016)
5. Barbosa de Santis, R., Azevedo Costa, M.: Extended isolation forests for fault detection in small hydroelectric plants. Sustainability **12**(16), 6421 (2020)
6. Selak, L., Butala, P., Sluga, A.: Condition monitoring and fault diagnostics for hydropower plants. Comput. Ind. **65**(6), 924–936 (2014)
7. Liu, Y., Xu, Y., Liu, J., Li, S., Cao, H., Chen, J.: Operational risk assessment for hydroelectric generating units using multi-head spatio-temporal attention and adaptive fuzzy clustering. Meas. Sci. Technol. **35**(2), 025011 (2024)
8. Spasenic, Z., Makajic-Nikolic, D., Benkovic, S.: Integrated FTA-risk matrix model for risk analysis of a mini hydropower plant's project finance. Energy Sustain. Dev. **70**, 511–523 (2022)
9. Shaktawat, A., Vadhera, S.: Risk management of hydropower projects for sustainable development: a review. Environ. Dev. Sustain. **23**(1), 45–76 (2021)
10. Yu, C., Wang, Z., Fang, R., Huang, D., Huang, T., Zhang, W.: Shutdown risk assessment of small-size hydropower station under typhoon disaster. In: Conference on Energy Internet and Energy System Integration, pp. 1756–1761. IEEE (2020)
11. Spasenic, Z., Makajic-Nikolic, D., Benkovic, S.: Risk assessment of financing renewable energy projects: a case study of financing a small hydropower plant project in Serbia. Energy Rep. **8**, 8437–8450 (2022)
12. Pinciroli, R., et al.: Epistemic uncertainty propagation in a weibull environment for a two-core system-on-chip. In: 2017 2nd International Conference on System Reliability and Safety (ICSRS), pp. 516–520 (2017)
13. Trivedi, K.S., Bobbio, A.: Reliability and Availability Engineering. Analysis and Applications. Cambridge University Press, Modeling (2017)

# On Portability of Software Performance Solutions to the Energy Domain

Vittorio Cortellessa[(✉)][iD]

Department of Information Engineering, Computer Science and Mathematics,
University of L'Aquila, L'Aquila, Italy
`vittorio.cortellessa@univaq.it`

**Abstract.** The evolution of the software performance research domain has gone through several fundamental steps that represent the milestones of this discipline. In this paper, five of these steps at the modeling and code level are considered and, after having motivated their relevance in the software performance domain, it is envisioned whether they could be ported and applied to the energy consumption domain, given the natural proximity of performance and energy in software/hardware systems.

**Keywords:** Software Performance · Software Modeling · Code analysis and testing · Energy Consumption

## 1 Introduction

The exponential growth of software-controlled devices, along with the widespread of AI-based applications, claim for amounts of energy so high that even huge investments cannot guarantee in the near future. Therefore, it is nowadays well-known that novel approaches are necessary in software/hardware design to lower the energy requirements.

A lot of effort has been undeniably spent in the last decade, on one side, to design low-energy hardware devices and, on the other side, to introduce innovative techniques in the software development lifecycle aimed at targeting energy-aware software that, independently on the underlying layers, exploits at the best its context for sake of energy saving [11].

Software performance and energy consumption are intrinsically related, mostly due to the fact that both depend on the way software uses the underlying resources (i.e., CPU, memory, network bandwith, etc.). Several studies have been presented to analyze the relationship between software performance and energy [19,35].

In this paper, one preliminary and five breakthroughing steps are described in the software performance research domain with the aim of conceptually analyzing whether they can be ported or (at least) any similar effort has been spent in the energy consumption domain.

The paper is organized as follows: in Sect. 2 a preliminary step about the separation of concerns between software and hardware is introduced; Sect. 3 describes

© The Author(s), under exclusive license to Springer Nature Switzerland AG 2026
L. Carnevali and J. Doncel (Eds.): EPEW 2025, LNCS 15657, pp. 29–40, 2026.
https://doi.org/10.1007/978-3-032-16345-5_3

three steps at the modeling level, whereas Sect. 4 presents two steps at the code level; in Sect. 5 some critical issues are discussed; finally, Sect. 6 concludes the paper.

## 2   A Separation of Concerns

The research area on performance analysis of computing systems arised in the '70 s, when the first performance models appeared [21]. Stochastic modeling languages, like Queueing Networks and Petri Nets, were already available, even though the complexity of solution processes has always limited the scalability of those models.

A specific aspect that has characterized performance models for several years was that software and hardware parameters were not clearly distinguishable in the model. As an example, let us consider the service time $S$ of a Queueing Network node that typically represents a device. It is well-known that this parameter has to combine the intrinsic speed $\mu$ of the device (e.g., expected number of assembler operations per unit of time in a CPU) with the amount $d$ of resource demanded by the software job (e.g., number of assembler operations to complete the job). Service time is obtained as $S = d/\mu$.

Indeed, the demand $d$ has started to assume the role of first-class citizen in performance modeling around mid '80 s, when the concept of number of visits has been introduced [22]. This has enabled a clear separation of concerns between hardware *platform-specific* parameters and software *platform-independent* ones. A straightforward consequence has been the major focus given to software (static and behavioral) aspects in performance modeling.

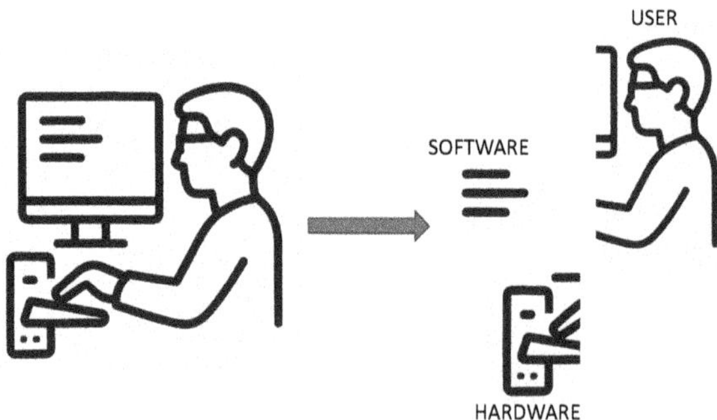


**Fig. 1.** Separation of hardware and software concerns.

Figure 1 graphically illustrates this fundamental step that has opened to several previously hidden possibilities, such as: (i) characterizing the user workload on different software use cases; (ii) easily constructing the models of the same software system on different hardware platforms, and viceversa; (iii) more knobs in the hands of performance experts to solve performance problems emerging from model solutions (i.e., beyond bottleneck device identification), such as software refactoring/redeploying.

**In the Energy Domain.** An analogous separation of concerns is necessary when modeling energy consumption of software/hardware systems. An interesting contribution in this direction has been given in [4], where it is shown that automatic random test generation with resource utilization heuristics can be used successfully to build accurate software energy consumption models. Indeed, resource utilization is a key concept in any energy model that intends to keep separation between software and hardware parameters. For example, the abstract amount of CPU that a line of code needs to be completed is independent on the amount of energy that the CPU spends. Hence, by keeping distinction between those two parameters it is possible to build more testing scenarios and, consequently, a wider set of repairing actions for energy problems that may emerge from testing.

## 3   At Modeling Level

This section deals with three major steps at the modeling level, each treated in a different subsection.

### 3.1   Automated Model Generation

The rapid evolution of Model-Driven Engineering at the end of '90 s has heavily affected even the software performance domain. Indeed, the possibility of representing software/hardware models with advanced multi-view modeling languages (such as UML) has enabled the automated generation of performance models from software artifacts.

Figure 2 shows a generic performance model generation process from a software model, which has been instantiated in many contexts in the first decade of the 2000s [1,2,7,15]. The most relevant points that enable such a process are: (i) the source software model is *Performance-annotated*, namely performance parameters have to be additionally annotated in the software model; (ii) a model-to-model transformation (i.e., SAM2PM in the figure) has to be coded and implemented to guarantee that the semantics of the target performance model is compliant with the source software model one.

**In the Energy Domain.** An interesting aspect that should be studied is whether an add-on to this process would be sufficient for enabling it to work in the energy domain. In other words, would additional annotations in the original software model be sufficient for treating with energy aspects? Some approaches were appeared few years ago with the intent of extending existing modeling languages to the energy domain [18]. As a consequence to the potential extension of modeling languages, would an increment to the model transformation be sufficient for embedding those annotations in the process, so to obtain a target model that deals with software energy consumption?

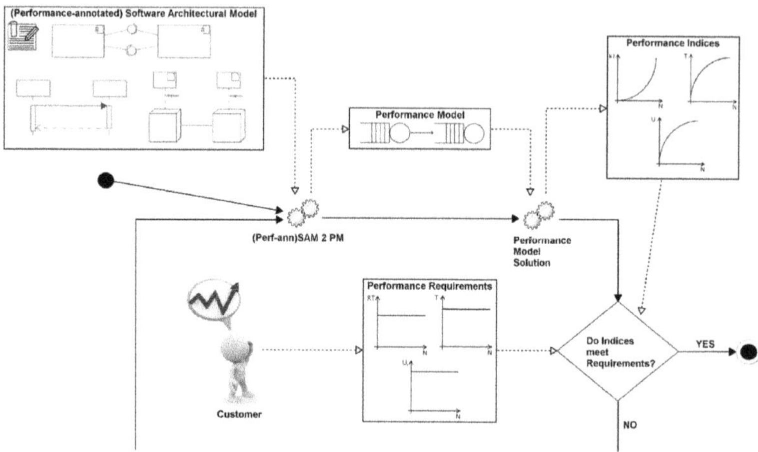


**Fig. 2.** Automated model generation process.

## 3.2   Result Interpretation: Performance Antipatterns

Figure 2 is evidently incomplete, because the arrow departing from performance requirement violation is hanging. Figure 3 shows a whole round-trip software performance engineering process, where a backward path appears below the forward path illustrated in the previous subsection. The backward path is in charge of interpreting the results of performance model solutions and proposing design solutions that overcome emerged problems.

Specifically, the backward path in this case is based on the concept of performance antipattern, which is a bad practice in software modeling that usually leads to performance degradation. In literature, a certain number of performance antipatterns have been first informally described [33], and thereafter formalized [5]. Formalization has allowed, as illustrated in Fig. 3, to start from a database of rules (PAs) that can feed an engine detecting possible antipatterns on the model under analysis. Formalization has also sustained the search of antipattern solutions in terms of model refactoring actions (RAs(PAs)). By applying these actions on the original model, antipatterns shall disappear and the performance shall consequently improve [6].

**In the Energy Domain.** This is probably the most promising step to be straightforwardly ported to the energy domain. Indeed, it is quite intuitive that there exist practices in the software development lifecycle that can badly affect the energy consumption of a software system. A recent work has indeed appeared on the classification of energy patterns for web [30], where the authors aim to understand whether good practices can be applied for sake of energy saving in the web. It would be valuable to contribute in this direction towards antipatterns identification and formalization, and the knowledge acquired in the performance domain could facilitate the path.

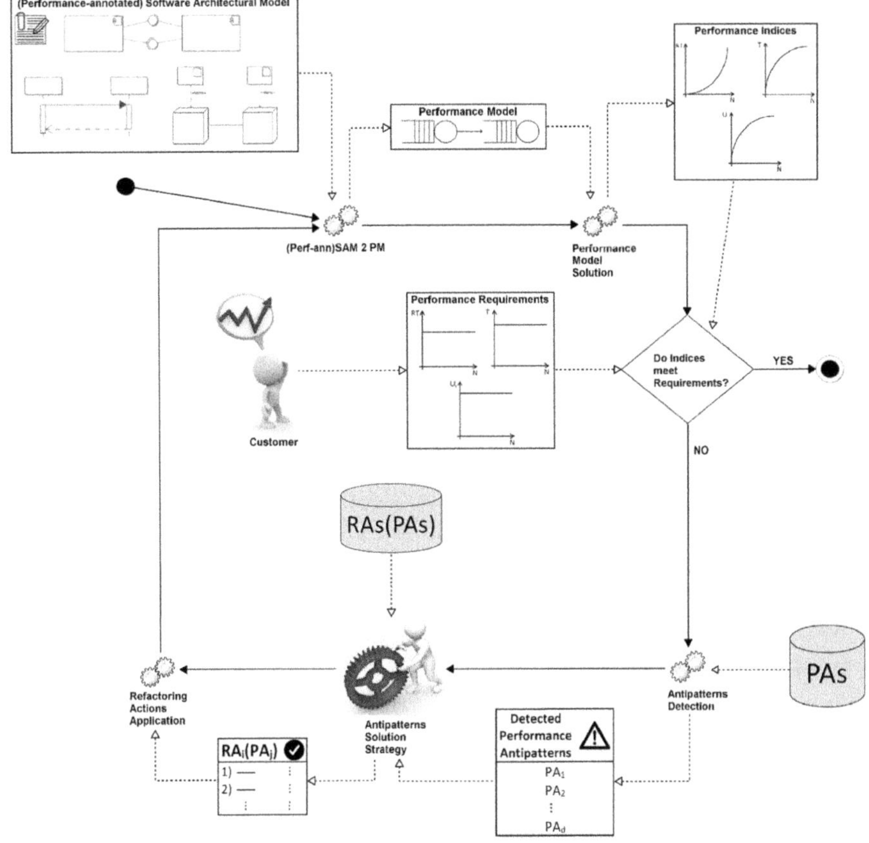


**Fig. 3.** Round-trip software performance engineering process.

## 3.3   Searching Solutions: Multi-objective Optimization

The idea of software refactoring for performance improvement can be generalized
in a process aimed at continuously searching optimal solution with respect to
a certain number of non-functional attributes. Indeed, in the last two decades,
several approaches have appeared that exploit searching techniques adopted in
multi-objective optimization (e.g., genetic algorithms) for finding Pareto-optimal
solutions with respect to a certain number of objectives. Obviously, the refactoring actions applied to an initial model must preserve the functional equivalence
of the generated models.

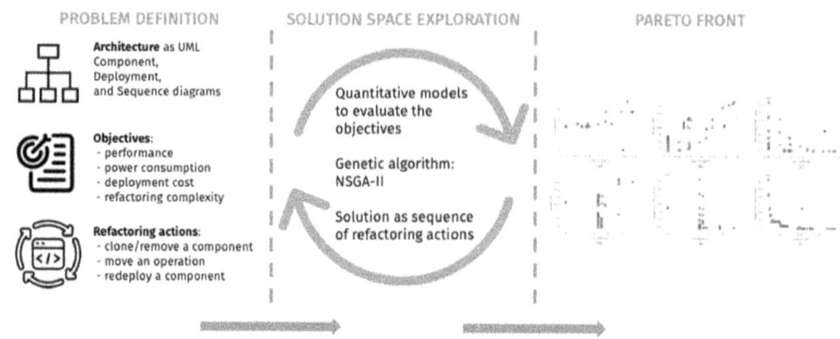


**Fig. 4.** An instance of multi-objective modeling for non-functional optimization.

Figure 4 summarizes an instance of multi-objective optimization in the domain of non-functional attributes. In particular, four objectives are defined on a software architecture represented by static and dynamic UML diagrams, namely: performance, power consumption, deployment cost and refactoring complexity [9]. The optimization engine here is based on NSGA-II [13], and the Pareto solutions represents other architectural models obtained by applying (semantic-preserving) refactoring actions on the original one.

**In the Energy Domain.** The example above already deals with energy issues in combination with other non-functional attributes. Multi-objective optimization has been widely used in the last two decades in many different contexts with the aim of finding near-optimal software solutions that satisfy quality aspects, including multi-energy systems in the field of vehicle recharging [27]. Energy consumption in the context of software/hardware systems can certainly enter this picture, at the only cost of developing energy models that can fit optimization solution algorithms.

## 4    At Code Level

Relevant progresses have been made in the last years on performance analysis at code level, mostly due to innovative techniques for inspection of software code and runtime monitoring. Initially, the attention had been given to the possibility of reconstructing performance models from runtime traces [17,26]. In the last decade, quite sophisticated approaches have been introduced for detecting performance regressions on running code [10,12,28,29], and for identifying root causes of regressions [14,36].

This section focuses on two specific issues related to the performance analysis at the code level, each treated in a different subsection.

### 4.1    Performance Prediction Based on Time Series Analysis

The recent availability of data collected on software at runtime has opened the possibility of conducting analyses on data time series. At the same time, advanced statistical techniques, as well as AI-based ones, have enabled sophisticated approaches to time series analysis, as illustrated in Fig. 5.

In particular, time series forecasting (TSF) offers unique opportunities to enhance software monitoring and facilitate proactive issue resolution. Indeed, the possibility of forecasting the trend of runtime metrics (e.g., execution time, crash probability, etc.) allows to activate countermeasures in due time to prevent unrepairable failures.

An interesting aspect of this research area is that different techniques have demonstrated to work better on certain types of metrics instead of other ones, and also that their effectiveness can vary depending on the required forecasting horizon [25]. For example, neural network-based techniques can work better on execution times than on number of crashes, but their effectiveness vanishes when the forecasting offset becomes too long.

**In the Energy Domain.** Time series forecasting has been widely studied in the last few years for sake of predicting the energy consumption of computer systems [23, 31], as well as other contexts like urbanization [3]. Given the promising results that it has been demonstrated in adjacent contexts, the potential of learning should be deeply investigated for computer energy issues.

## 4.2   Warmup vs Steady State

Time series often present some instability in the metric values at the beginning, while stabilizing their values after some time. This is especially true in systems (like software/hardware ones) that usually incur in startup operations based on the system status (e.g., just-in-time compilation). It is therefore very important to identify a changepoint in a time series that separates the initial unstable part, namely "warmup", from the following more stable part, namely "steady state". Datapoints collected in the warmup phase are obviously discarded by the time series analysis technique.

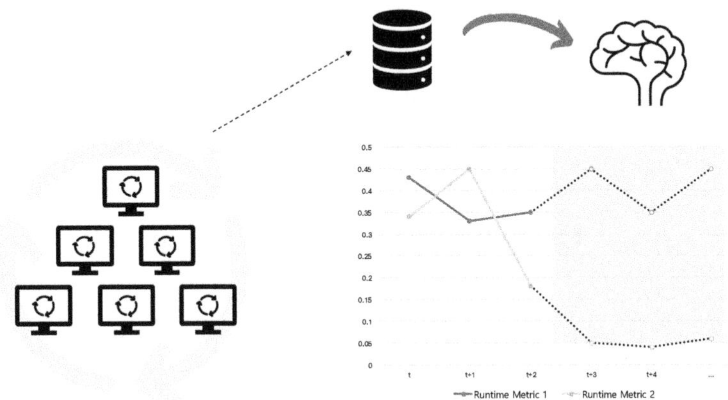


**Fig. 5.** Software metrics time series forecasting.

Of course, time series can present some instability and outlier even after warmup time, but the identification of a reliable changepoint is a critical step for collecting high quality datapoints. Indeed, as illustrated in Fig. 6, a too early identification of changepoint would lead to consider unstable data in the analysis, whereas a too late identification would lead to loose good quality data and, as a consequence, to spend more time than needed to collect a sufficient amount of datapoints.

It has been shown, in the domain of performance testing, that this problem can be promisingly tackled by using AI-based techniques [34]. The idea is to train a model with a ground truth set built on human observation of time series, and thereafter using the model on field to identify the end of warmup phase on unseen time series.

**In the Energy Domain.** The collection of metrics in the energy domain could be subject to a similar phenomenon, in that the underlying systems can be unstable for a certain initial period of time. Hence, it seems to be worth to investigate warmup time of energy metrics in computer systems to improve the quality of collected data.

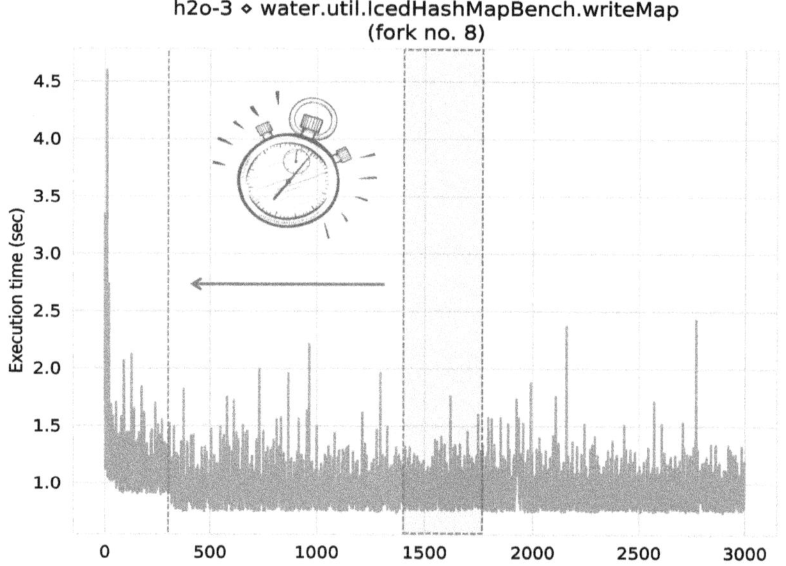


**Fig. 6.** The warmup problem in timeseries.

## 5   Critical Issues

As mentioned in Sect. 4, an early objective of research at the code level has been to reconstruct performance models from running software systems. Being performance a non-functional properties that emerges at runtime, it is well-assessed that the code structure does not help in this task, as also demonstrated by recent studies based on learning techniques that have produced negative results [16]. Therefore, it has been clear from the beginning that code had to be instrumented for trying to capture its dynamic behavior at runtime.

In practice, the collection of performance metrics and software traces at runtime, as discussed in the previous section, not only may help to reconstruct performance models, but it may serve to many other goals like time series forecasting, root cause analysis of performance degradation, and optimal task allocation. Similar considerations can be made in the energy domain, where software/hardware behavior is essential for the analysis, and where accurate energy metrics measurement could support multiple design and runtime decisions.

However, runtime measurements in both domains (i.e., performance and energy) incur in a critical issue represented by instrumentation overhead. Although several software tools and benchmarks have been developed for collecting runtime software metrics [32], software code available on common repositories still needs to undergo an instrumentation phase before opening a runtime collection campaign. Apart from being a time-consuming task, on one side this task needs to be accurate in the identification of code sections of interest, and on the other side it has to be minimally invasive in terms of computation time and/or energy required to not compromising the result reliability. Therefore, this remains an aspect that needs to be still investigated in the near future. In addition, the energy domain presents other two critical issues to address, that are measurement noise [20] and interactions with dynamic voltage/frequency scaling [24].

Accurate measurements at runtime also contribute to closing the loop with high-level models. Indeed, metrics can help to understand whether the high-level models need some tuning to reflect the real behavior of a software/hardware system [8], and this holds for both cases of performance and energy. The accuracy of models mostly depends on their ability to capture the essential aspects of a system. This can be quite complex in cases where hidden layers (such as middleware, operating systems, etc.) are prevalent in their contribution to the target metrics. Indeed, such layers are usually taken into account in high-level models by allocating a fixed amount or percentage of resources to them. Where this assumption is not realistic (e..g., where the operating system needs to manage tasks with high variance of resource requirements), the roundtrip between monitoring and modeling is hard to achieve.

## 6   Conclusion

In this paper, some fundamental software performance solutions have been illustrated, with the aim of envisaging their porting to the energy consumption domain. In particular, three aspects at the level of modeling and two aspects at the level of code have been considered, with the aim of providing some seed for further investigation o these aspects in the energy domain. Since performance and energy are strictly dependent in terms of their metrics, this work feeds the idea that those techniques that have been adopted in the performance domain to address critical problems could be equally successful if applied in the energy domain.

# References


1. Andolfi, F., Aquilani, F., Balsamo, S., Inverardi, P.: Deriving performance models of software architectures from message sequence charts. In: Second International Workshop on Software and Performance, WOSP 2000, Ottawa, Canada, 17–20 September 2000, pp. 47–57. ACM (2000)
2. Arief, L.B., Speirs, N.A.: A UML tool for an automatic generation of simulation programs. In: Second International Workshop on Software and Performance, WOSP 2000, Ottawa, Canada, 17–20 September 2000, pp. 71–76. ACM (2000)
3. Chou, J.S., Tran, D.S.: Forecasting energy consumption time series using machine learning techniques based on usage patterns of residential householders. Energy **165**, 709–726 (2018)
4. Chowdhury, S.A., Borle, S., Romansky, S., Hindle, A.: Greenscaler: training software energy models with automatic test generation. Empir. Softw. Eng. **24**(4), 1649–1692 (2019)
5. Cortellessa, V., Di Marco, A., Trubiani, C.: An approach for modeling and detecting software performance antipatterns based on first-order logics. Softw. Syst. Model. **13**(1), 391–432 (2014)
6. Cortellessa, V., Di Marco, A., Trubiani, C.: An approach for modeling and detecting software performance antipatterns based on first-order logics. Softw. Syst. Model. **13**(1), 391–432 (2014)
7. Cortellessa, V., Mirandola, R.: PRIMA-UML: a performance validation incremental methodology on early UML diagrams. Sci. Comput. Program. **44**(1), 101–129 (2002)
8. Cortellessa, V., Di Pompeo, D., Eramo, R., Tucci, M.: A model-driven approach for continuous performance engineering in microservice-based systems. J. Syst. Softw. **183** (2022)
9. Cortellessa, V., Di Pompeo, D., Tucci, M.: Exploring sustainable alternatives for the deployment of microservices architectures in the cloud. In: 21st IEEE International Conference on Software Architecture, ICSA 2024, Hyderabad, India, 4–8 June 2024, pp. 34–45. IEEE (2024)
10. Cortellessa, V., Traini, L.: Detecting latency degradation patterns in service-based systems. In: Proceedings of the ACM/SPEC International Conference on Performance Engineering, pp. 161–172. ICPE '20, Association for Computing Machinery (2020)
11. D'Agostino, D., Merelli, I., Aldinucci, M., Cesini, D.: Hardware and software solutions for energy-efficient computing in scientific programming. Sci. Program. **2021**, 5514284:1–5514284:9 (2021)
12. Daly, D., Brown, W., Ingo, H., O'Leary, J., Bradford, D.: The use of change point detection to identify software performance regressions in a continuous integration

system. In: Proceedings of the ACM/SPEC International Conference on Performance Engineering, pp. 67–75. ICPE '20, Association for Computing Machinery (2020)

13. Deb, K., Agrawal, S., Pratap, A., Meyarivan, T.: A fast elitist non-dominated sorting genetic algorithm for multi-objective optimization: NSGA-II. In: Schoenauer, M., et al. (eds.) PPSN 2000. LNCS, vol. 1917, pp. 849–858. Springer, Heidelberg (2000). https://doi.org/10.1007/3-540-45356-3_83

14. Heger, C., Happe, J., Farahbod, R.: Automated root cause isolation of performance regressions during software development. In: Proceedings of the 4th ACM/SPEC International Conference on Performance Engineering, pp. 27–38. ICPE '13, Association for Computing Machinery (2013)

15. Hermanns, H., Jansen, D.N., Usenko, Y.S.: From stocharts to modest: a comparative reliability analysis of train radio communications. In: Proceedings of the Fifth International Workshop on Software and Performance, WOSP 2005, Palma, Illes Balears, Spain, 12–14 July 2005, pp. 13–23. ACM (2005)

16. Imran, M., Cortellessa, V., Di Ruscio, D., Rubei, R., Traini, L.: Is code coverage of performance tests related to source code features? An empirical study on open-source java systems. Empir. Softw. Eng. **30**(157) (2025). https://doi.org/10.1007/s10664-025-10712-3

17. Israr, T.A., Lau, D.H., Franks, G., Woodside, C.M.: Automatic generation of layered queuing software performance models from commonly available traces. In: Proceedings of the Fifth International Workshop on Software and Performance, WOSP 2005, Palma, Illes Balears, Spain, 12–14 July 2005, pp. 147–158. ACM (2005)

18. Kang, E., Perrouin, G., Schobbens, P.: Towards formal energy and time aware behaviors in EAST-ADL: an MDE approach. In: Tang, A., Muccini, H. (eds.) 2012 12th International Conference on Quality Software, Xi'an, Shaanxi, China, 27–29 August 2012, pp. 124–127. IEEE (2012)

19. Khan, M.A., Hankendi, C., Coskun, A.K., Herbordt, M.C.: Software optimization for performance, energy, and thermal distribution: initial case studies. In: 2011 International Green Computing Conference and Workshops, IGCC 2012, Orlando, FL, USA, 25–28 July 2011, pp. 1–6. IEEE Computer Society (2011)

20. Langdon, W.B., Petke, J., Bruce, B.R.: Optimising quantisation noise in energy measurement. In: Handl, J., Hart, E., Lewis, P.R., López-Ibáñez, M., Ochoa, G., Paechter, B. (eds.) PPSN 2016. LNCS, vol. 9921, pp. 249–259. Springer, Cham (2016). https://doi.org/10.1007/978-3-319-45823-6_23

21. Lavenberg, S.S.: Queueing analysis of a multiprogrammed computer system having a multilevel storage hierarchy. SIAM J. Comput. **2**(4), 232–252 (1973)

22. Lazowska, E.: Quantitative System Performance: Computer System Analysis Using Queueing Network Models. Prentice-Hall, Hoboken (1984)

23. Liu, X., et al.: A data mining research on office building energy pattern based on time-series energy consumption data. Energy Buildings **259** (2022)

24. Maioli, A., Quinones, K.A., Ahmed, S., Alizai, M.H., Mottola, L.: Dynamic voltage and frequency scaling for intermittent computing. ACM Trans. Sen. Netw. **21**(2) (2025)

25. Di Menna, F., Traini, L., Cortellessa, V.: Time series forecasting of runtime software metrics: an empirical study. In: Balsamo, S., Knottenbelt, W.J., Abad, C.L., Shang, W. (eds.) Proceedings of the 15th ACM/SPEC International Conference on Performance Engineering, ICPE 2024, London, United Kingdom, 7–11 May 2024, pp. 48–59. ACM (2024)

26. Mizan, A., Franks, G.: An automatic trace based performance evaluation model building for parallel distributed systems. In: Kounev, S., Cortellessa, V., Mirandola, R., Lilja, D.J. (eds.) ICPE'11 - Second Joint WOSP/SIPEW International Conference on Performance Engineering, Karlsruhe, Germany, 14–16 March 2011, pp. 61–72. ACM (2011)
27. Murray, P., Carmeliet, J., Orehounig, K.: Multi-objective optimisation of power-to-mobility in decentralised multi-energy systems. Energy **205**, 117792 (2020)
28. Nguyen, T.H., Adams, B., Jiang, Z.M., Hassan, A.E., Nasser, M., Flora, P.: Automated detection of performance regressions using statistical process control techniques. In: Proceedings of the ACM/SPEC International Conference on Performance Engineering, pp. 299–310. ICPE '12, Association for Computing Machinery (2012)
29. Ramakrishnan, R., Kaur, A.: Technique for detecting early-warning signals of performance deterioration in large scale software systems. In: Proceedings of the 8th ACM/SPEC on International Conference on Performance Engineering, pp. 213–222. ICPE '17, Association for Computing Machinery (2017)
30. Rani, P., Zellweger, J., Kousadianos, V., Cruz, L., Kehrer, T., Bacchelli, A.: Energy patterns for web: an exploratory study. In: Proceedings of the 46th International Conference on Software Engineering: Software Engineering in Society, ICSE-SEIS2024, Lisbon, Portugal, 14–20 April 2024, pp. 12–22. ACM (2024)
31. Romansky, S., Borle, N.C., Chowdhury, S., Hindle, A., Greiner, R.: Deep green: modelling time-series of software energy consumption. In: 2017 IEEE International Conference on Software Maintenance and Evolution (ICSME), pp. 273–283 (2017)
32. Schmitt, N., Lange, K., Sharma, S., Rawtani, N., Ponder, C., Kounev, S.: The specpowernext benchmark suite, its implementation and new workloads from a developer's perspective. In: Bourcier, J., Jiang, Z.M.J., Bezemer, C., Cortellessa, V., Pompeo, D.D., Varbanescu, A.L. (eds.) ICPE '21: ACM/SPEC International Conference on Performance Engineering, Virtual Event, France, 19–21 April 2021, pp. 225–232. ACM (2021)
33. Smith, C.U., Williams, L.G.: Software performance antipatterns. In: Second International Workshop on Software and Performance, WOSP 2000, Ottawa, Canada, 17–20 September 2000, pp. 127–136. ACM (2000)
34. Traini, L., Di Menna, F., Cortellessa, V.: AI-driven java performance testing: balancing result quality with testing time. In: Filkov, V., Ray, B., Zhou, M. (eds.) Proceedings of the 39th IEEE/ACM International Conference on Automated Software Engineering, ASE 2024, Sacramento, CA, USA, October 27 - November 1 2024, pp. 443–454. ACM (2024)
35. Tüysüz, M.F., Ankarali, Z.K., Gözüpek, D.: A survey on energy efficiency in software defined networks. Comput. Netw. **113**, 188–204 (2017)
36. Zhao, Y., et al.: How are performance issues caused and resolved?-An empirical study from a design perspective. In: Proceedings of the ACM/SPEC International Conference on Performance Engineering, pp. 181–192. ICPE '20, Association for Computing Machinery (2020)

# Accelerating Mobile Inference Through Fine-Grained CPU-GPU Co-Execution

Zhuojin Li, Marco Paolieri$^{(\boxtimes)}$, and Leana Golubchik

University of Southern California, Los Angeles, CA, USA
`{zhuojinl,paolieri,leana}@usc.edu`

**Abstract.** Deploying deep neural networks on mobile devices is increasingly important but remains challenging due to limited computing resources. On the other hand, their unified memory architecture and narrower gap between CPU and GPU performance provide an opportunity to reduce inference latency by assigning tasks to both CPU and GPU. The main obstacles for such collaborative execution are the significant synchronization overhead required to combine partial results, and the difficulty of predicting execution times of tasks assigned to CPU and GPU (due to the dynamic selection of implementations and parallelism level). To overcome these obstacles, we propose both a lightweight synchronization mechanism based on OpenCL fine-grained shared virtual memory (SVM) and machine learning models to accurately predict execution times. Notably, these models capture the performance characteristics of GPU kernels and account for their dispatch times. A comprehensive evaluation on four mobile platforms shows that our approach can quickly select CPU-GPU co-execution strategies achieving up to 1.89x speedup for linear layers and 1.75x speedup for convolutional layers (close to the achievable maximum values of 2.01x and 1.87x, respectively, found by exhaustive grid search on a Pixel 5 smartphone).

**Keywords:** Mobile Inference · Co-Execution · Latency · Prediction

## 1 Introduction

Machine learning (ML) techniques have achieved rapid growth in recent years, driven by the breakthroughs in large-scale architectures such as Large Language Models (LLMs) and Large Vision Models (LVMs). These state-of-the-art models exhibit impressive capabilities in a wide range of applications, including question answering, image recognition, and video analysis. While these models are typically trained on powerful cloud servers, there is a growing demand for deployment on mobile platforms to serve inference tasks, since on-device deployment not only protects user privacy, but also provides offline availability and reduces latency for real-time applications.

However, deploying large-scale models on mobile platforms remains challenging, due to limited computational resources and energy constraints. To improve


L. Carnevali and J. Doncel (Eds.): EPEW 2025, LNCS 15657, pp. 41–55, 2026.
https://doi.org/10.1007/978-3-032-16345-5_4

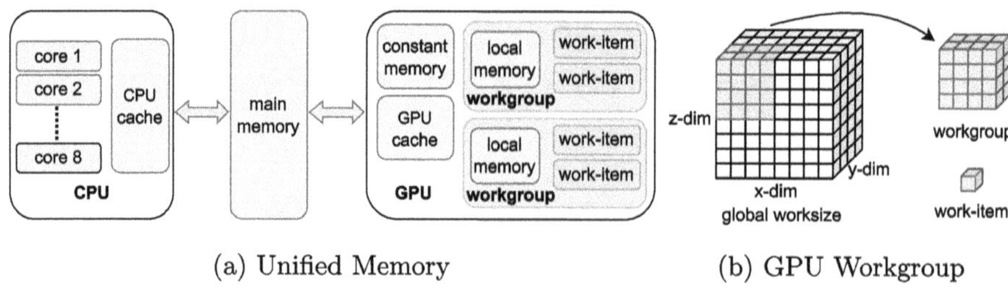



(a) Unified Memory    (b) GPU Workgroup

**Fig. 1.** Introduction to Mobile Platforms

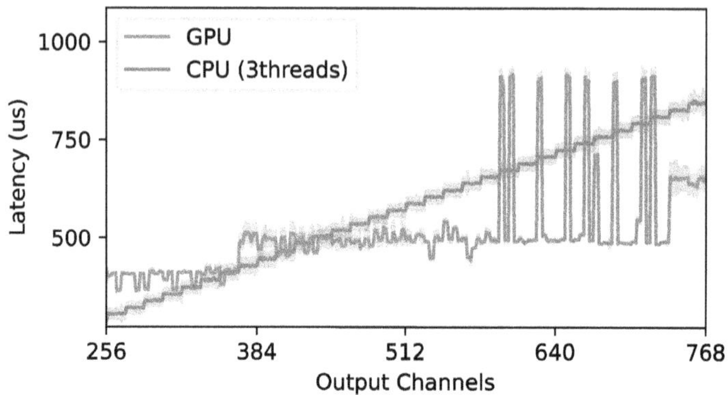



**Fig. 2.** Comparison of CPU and GPU Latencies for Linear Operations with Input Shape (50, 3072), with 95% Confidence Intervals (OnePlus 11)

inference latency on mobile platforms, substantial research and industrial efforts have focused on developing efficient neural network architectures [16] and specialized hardware [7]. On top of these efforts, our work explores an additional dimension: distributing fine-grained operations (e.g., forward propagation of a neural network layer) across heterogeneous computing devices (specifically, CPU and GPU) available on modern mobile platforms. In particular, we pursue this approach by leveraging the following opportunities.

The unified memory architecture on mobile platforms enables both the main processor (CPU) and specialized accelerators (e.g., GPU and NPU) to directly access shared areas of main memory, as illustrated in Fig. 1a. In contrast, GPUs of cloud servers maintain separate on-chip memories and require data transfers to share data with the CPU through the main memory. By eliminating the need for memory transfers to main memory, unified memory on mobile platforms facilitates collaborative execution of inference tasks across compute devices.

Moreover, in contrast with existing works [8,19] which report poor performance on mobile CPUs (due to inefficient CPU implementations in some ML libraries such as OpenVINO [6]), our empirical analysis shows that the performance gap between mobile CPUs and GPUs can be narrow for important operations when using the XNNPACK [4] library in TensorFlow Lite (TFLite), which

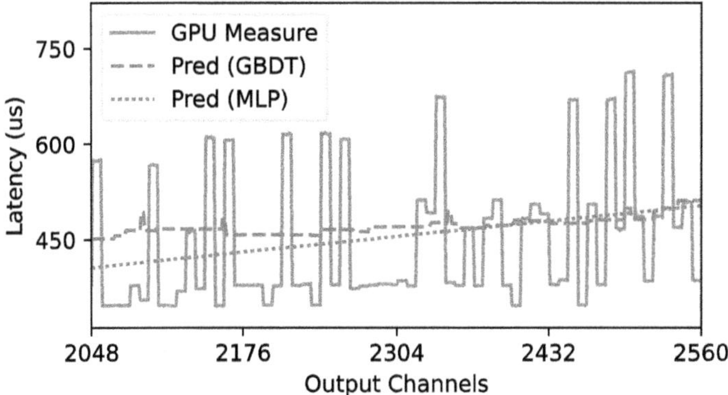


**Fig. 3.** Inadequate Modeling of GPU Latency Spikes by Existing Methods for Linear Operations with Input Shape (50, 768) (OnePlus 11)

provides high-performance implementations based on advanced SIMD instructions for ARM CPUs. For instance, as illustrated in Fig. 2, for matrix multiplications (i.e., a linear layer in a neural network) of sizes $50 \times 3072$ and $3072 \times C_{out}$, a CPU implementation with 3 threads achieves even lower latency than a GPU kernel when $C_{out} < 425$. The significant performance of mobile CPUs motivates us to *assign compute tasks to both CPU and GPU to achieve lower inference latency through parallel execution.*

However, designing an effective co-execution strategy across multiple accelerators remains challenging. In our empirical evaluation, existing approaches achieve limited performance gains mainly due to the following two reasons.

First, optimal workload partitioning across CPUs and GPUs is difficult, primarily because GPU kernels exhibit complex non-linear performance characteristics. Our experiments, reported in Fig. 3, show significant latency spikes (green line) for linear operations as the output dimension $C_{out}$ increases, due to heuristic choices in GPU kernel implementations and in the assignment of tasks to *workgroups* (i.e., to groups of threads executing the same code on different subsets of the input data, as illustrated in Fig. 1b). Consequently, co-execution frameworks relying on linear models for GPU latency prediction (e.g., [2]) can make poor partitioning decisions. Although non-linear ML models based on operation configurations (e.g., input/output channels) have been proposed for more accurate latency predictions [9, 13, 15, 22], our evaluations in Fig. 3 indicate that these methods[1] still fail to capture sudden latency spikes for specific configurations.

Second, synchronization overhead between compute devices is often non-negligible. Existing work [9] reports overhead of up to 1 ms for the GPU to notify the CPU that data mapping of a shared memory region has completed; as a comparison, our measurements indicate that, using the GPU on the OnePlus 11

---

[1] GBDT hyperparameters are detailed in Sect. 5.2. The MLP predictor was selected by varying the number of layers (1 to 4), neurons at each layer (32, 64 or 128), dropout rate (0 to 0.5), learning rate ($10^{-5}$ to 0.1) and weight decay ($10^{-6}$ to $10^{-3}$).

smartphone, the longest linear operation of the ViT-Base-32 neural network [3] takes only 660 µs. Such overhead can completely wipe out the benefits of co-execution; thus, minimizing synchronization overhead is crucial.

Our main contributions to tackling these problems are as follows:

- We develop accurate latency predictors that use kernel implementation details and kernel dispatch behaviors from TFLite (Sect. 3). Using detailed information, our predictors accurately capture complex latency characteristics, including the discontinuity due to heuristic workgroup choices and kernel selection, enabling more effective workload partitioning decisions.
- To address significant synchronization overhead, we use fine-grained shared virtual memory in OpenCL to implement an efficient CPU-GPU synchronization mechanism on mobile platforms (Sect. 4). Our design reduces expensive data mapping operations required for cache coherence and avoids notification delay through active querying. As a result, synchronization overhead is substantially reduced, e.g., from 162 µs to 7 µs for linear operations on a Motorola Edge Plus 2022 smartphone.
- We create a comprehensive dataset consisting of latency measurements of 2,039 linear and 2,051 convolution operations executed on four mobile devices using co-execution strategies with 1 to 3 CPU threads and the GPU. The evaluation shows that our predictors can quickly identify co-execution strategies achieving speedups up to 1.89x for linear operations and 1.75x for convolution operations (on Pixel 5 smartphones); we show that speedups are comparable to those identified through brute-force exploration of co-execution strategies (Sect. 5).

## 2   Inference Workload Partitioning

*ML Operation Partitioning.* Linear and convolutional layers are fundamental building blocks in deep neural networks. A linear layer multiplies an input matrix $\mathbf{X} \in \mathbb{R}^{L \times C_{in}}$ (activations from the previous layer) by the *weights* $\mathbf{W} \in \mathbb{R}^{C_{in} \times C_{out}}$ (the trainable parameters of the layer); each column of the output $\mathbf{Y} = \mathbf{XW} \in \mathbb{R}^{L \times C_{out}}$ represents a different output feature. A convolutional layer is a generalization of a linear layer that is used to process data with grid-like topology (e.g., images): each output value $(\mathbf{Y})_{ijk}$ is obtained through the dot product of a *kernel* (or *filter*) $\mathbf{W}_k \in \mathbb{R}^{K \times K \times C_{in}}$ with a $K \times K$ patch of the input *feature map* $\mathbf{X} \in \mathbb{R}^{H_{in} \times W_{in} \times C_{in}}$ (2D activations from a previous layer), i.e., $(\mathbf{Y})_{ijk} = \mathbf{X}(i, j)\mathbf{W}_k$ where $\mathbf{X}(i, j)$ restricts the first two dimensions of $\mathbf{X}$ to $[i - \lfloor \frac{K}{2} \rfloor, i + \lfloor \frac{K}{2} \rfloor]$ and $[j - \lfloor \frac{K}{2} \rfloor, j + \lfloor \frac{K}{2} \rfloor]$, respectively. In the output $\mathbf{Y} \in \mathbb{R}^{H_{out} \times W_{out} \times C_{out}}$, input height and width can be reduced by using a *stride* $S > 1$ to increment $i$ and $j$, i.e., $H_{out} = \lfloor H_{in}/S \rfloor$ and $W_{out} = \lfloor W_{in}/S \rfloor$; the number of *output channels* $C_{out}$ is equal to the number of different kernels $\mathbf{W}_k$.

In this work, we focus on partitioning the computation of linear or convolutional layers along output channels. Since each output channel corresponds to a distinct column of a linear weight matrix $\mathbf{W}$ or to a distinct convolution

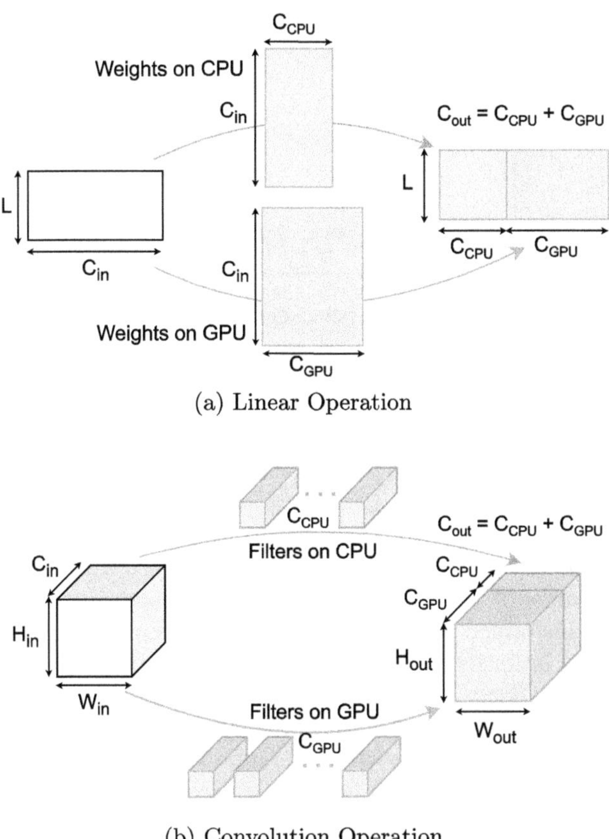


**Fig. 4.** Illustration of Computation Partitioning over Output Channels

kernel $\mathbf{W}_k$, each compute unit can store and manage its own subset of weights. Figure 4 illustrates this strategy: the total number of output channels $C_{out}$ is partitioned as $C_{CPU} + C_{GPU} = C_{out}$. Accordingly, the original weights are split by assigning the first $C_{CPU}$ columns of $\mathbf{W}$ or the first $C_{CPU}$ kernels $\mathbf{W}_k$ to the CPU and the rest to the GPU. During execution, CPU and GPU calculate their assigned portion of output independently using the shared input $\mathbf{X}$.

*Problem Formulation.* Formally, given a total of $C_{out}$ output channels for a linear or convolution operation, our goal is to determine the optimal partitioning $c_1 + c_2 = C_{out}$ that minimizes the parallel execution time, i.e.,

$$\min_{c_1 + c_2 = C_{out}} T_{overhead}(c_1, c_2) + \max\left(T_{CPU}(c_1), T_{GPU}(c_2)\right) .$$

Here, $T_{CPU}(c_1)$ and $T_{GPU}(c_2)$ represent the computation latency on CPU and GPU, respectively. $T_{overhead}(c_1, c_2)$ accounts for additional latency from synchronization overhead; in particular, we observe that the overhead remains con-

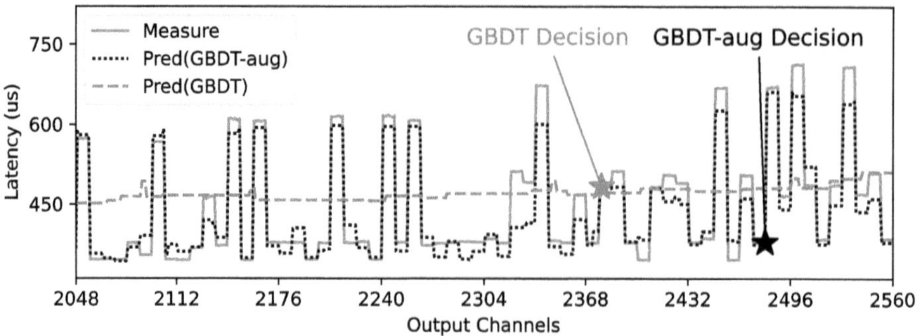


**Fig. 5.** Latency Prediction Improvement using Additional Features (OnePlus 11)

stant in our measurements when a co-execution strategy is used (Sect. 4), while $T_{overhead}(c_1, c_2) = 0$ when $c_1 = C_{out}$ or $c_2 = C_{out}$ (i.e., exclusive execution on CPU or GPU, respectively). The formulation aims at balancing computational loads across CPU and GPU, ensuring high resource utilization.

Since the number of output channels can be thousands in practice (e.g., 3,072 in the vision transformer ViT-Base-32 [3]) and the optimal partitioning varies across hardware platforms and other operation configurations (e.g., convolution kernel size), exhaustively measuring latency for every possible partitioning is costly. Thus, existing works [9, 11, 20] typically use ML-based latency predictors that rely on the operation parameters (e.g., input/output dimensions). However, as detailed in Sect. 3.1, these approaches can hardly capture the complex characteristics of GPU performance, resulting in suboptimal partitioning decisions. Additionally, as presented in Sect. 4, synchronization overhead $T_{overhead}$ must be accounted for when evaluating the achieved speedup, since significant overhead can diminish performance gains achieved through co-execution.

## 3   Accurate Latency Prediction

In this section, we present our approach to improve latency predictors used for partitioning decisions.

### 3.1   GPU Kernel Characterization

As motivated in Sect. 2, accurate latency prediction is crucial for effective partitioning decisions. To illustrate this, we conduct an experiment on a linear operation $\mathbf{W} \in \mathbb{R}^{768 \times 3072}$ from the vision transformer ViT-Base-32 [3], where it is used to transform an input feature map $\mathbf{X} \in \mathbb{R}^{50 \times 768}$ into $\mathbf{Y} \in \mathbb{R}^{50 \times 3072}$. To partition this operation, we measure the GPU latency of linear operations with size $768 \times C_{out}$ for $C_{out} \in [2048, 2560]$. As depicted in Fig. 5, GPU latency exhibits significant spikes, e.g., the linear operation with $C_{out} = 2500$ is counterintuitively 1.85 times slower than the one with $C_{out} = 2520$. Notably, a gradient-boosted

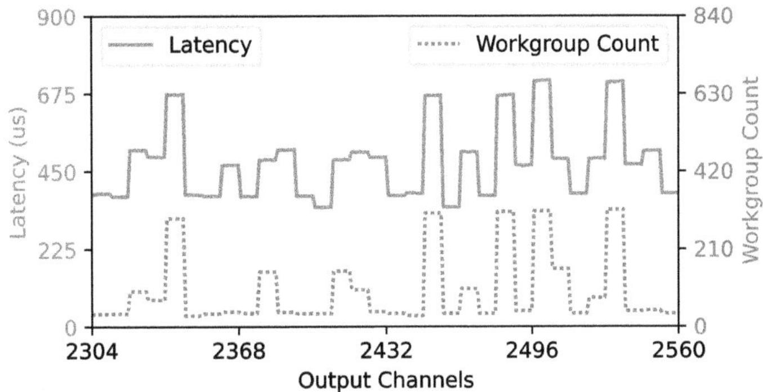


(a) Heuristic Workgroup for Linear Operations with Input Shape (50, 768)

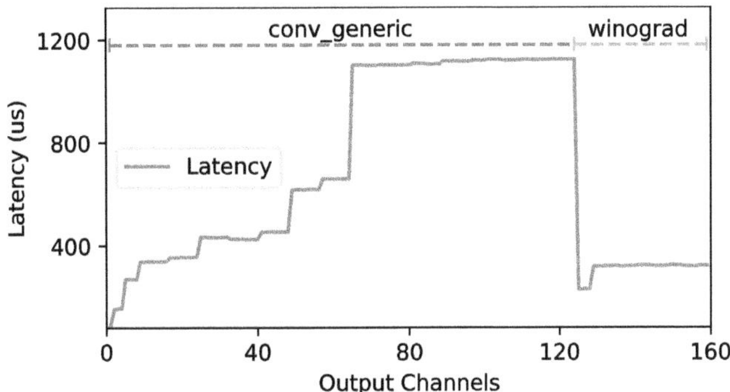


(b) Kernel Switch for 3x3 Convolution Operations with Input Shape (64, 64, 128)

**Fig. 6.** Reasons for Discontinuity in Latency Curve on Mobile GPUs (OnePlus 11)

decision tree (GBDT, an ML model broadly used in prior work [9,13,15,22]) using matrix sizes as input features only captures the overall increasing trend of latency (blue curve) rather than its sudden changes; using GBDT latency predictions, 2,378 output channels are assigned to the GPU (measured latency of 483 μs) and 694 channels are assigned to the CPU (measured latency of 424 μs), achieving only 1.02x speedup with respect to using only the GPU (as a reference, our approach in Sect. 3.2 achieves 1.29x speedup by assigning 2,480 channels to the GPU, based on more accurate predictions). To understand the discontinuity in GPU performance, we analyzed the source code of the ML framework TFLite [12] and identified two primary factors accounting for sudden latency changes:

1. *Heuristic Workgroup Choices.* In GPU computing, a workgroup is a collection of threads (work items) concurrently executing the same code and sharing resources such as local memory; the workgroup size determines how GPU threads are grouped for scheduling and is crucial for the efficiency of a GPU kernel. ML frameworks typically decide the workgroup size heuristically based

on the kernel implementation and on the underlying hardware architecture. Figure 6a presents the latencies and workgroup counts for linear operations with input size $50 \times 768$ and varying number of output channels $C_{out}$; here, we observe a strong correlation between the number of workgroups and kernel latency.

2. *Kernel Selection.* In order to enhance performance, TFLite provides multiple GPU kernel implementations of convolution operations for different parameters (kernel size/stride/channels, input/output size). For example, Fig. 6b shows that, for a convolutional layer with $3 \times 3$ filter and input size of $64 \times 64 \times 128$, when the number of output channels exceeds 128, the kernel implementation will switch to the Winograd algorithm, which offers higher efficiency for operations with more output channels. This change of kernel implementations results in substantial latency anomalies due to their distinct performance characteristics.

## 3.2   Feature Augmentation

To accurately predict the GPU latency, we propose a white-box approach to capture the two aforementioned factors. Specifically, we analyzed the algorithms of TFLite to determine kernel implementation and workgroup size for a given operation configuration. Below, we summarize our analysis.

First, TFLite mainly uses three kernel implementations for convolution operations: (1) `conv_constant`, which leverages the faster on-chip constant memory when sufficient registers (estimated based on output channels) are available and convolution filters can fit within the constant memory, (2) `winograd`, which reduces the number of multiplications when the filter size is small and input sizes are large enough to make the transformation overhead negligible compared to the savings in multiplications, and (3) `conv_generic`, which is the default implementation for general use cases. Second, once a kernel implementation is selected, ML frameworks heuristically select workgroup configurations by considering hardware-specific factors, such as number of registers, compute unit occupancy, and memory access patterns.

To leverage this additional information about GPU kernels, we (1) construct separate latency predictors for each kernel implementation, and (2) augment the predictor input features to include kernel dispatch information, such as size and number of workgroups; these dispatch-related features can be calculated based on the hardware specification and on the parameters of the operation being executed. Figure 5 illustrates that our approach (black curve) accurately captures the spikes of inference latency through feature augmentation. As a result, our enhanced latency prediction enables more effective workload partitioning for co-execution: using 2,480 output channels on GPU (with measured latency of 379 μs and prediction improved from 481 μs to 375 μs) and 592 output channels on CPU (with measured latency of 354 μs), the speedup improves from 1.02x to 1.29x (i.e., from 495 μs to 393 μs).

# 4   Reduction of CPU-GPU Synchronization Overhead

As discussed in Sect. 2, significant synchronization overhead can diminish the performance gains of co-execution. We identify two main sources of overhead.

1. *Data Mapping for Cache Coherence.* When CPU and GPU collaboratively compute the output of a neural network layer, data transfers between their memory hierarchies are necessary. In particular, since mobile CPU and GPU have distinct caches, explicit data mapping operations are required to maintain cache coherence. Previous works [9] report up to 1 ms data mapping overhead.
2. *Inter-Processor Notifications.* Since CPU and GPU computations proceed in parallel, synchronization points are required to manage data dependencies and enforce execution order. For example, in OpenCL, synchronization between the host (CPU) and device (GPU) can be implemented by making the GPU kernel dependent on a user event; once the CPU finishes its computation, it marks this event as completed. However, there is a delay before the GPU recognizes the updated event state. Similarly to earlier work [9], our measurements indicate that the delay is on average 162 μs on Motorola Edge Plus 2022 across 2,039 linear operations, which accounts for 39.9% of the total co-execution latency.

To reduce synchronization overhead, we use two techniques:

1. During inference, we store layer outputs in OpenCL *fine-grained shared virtual memory* (SVM), allowing both the OpenCL host (CPU) and device (GPU) to read and write directly to the same region in main memory. For instance, when a previous layer is evaluated collaboratively by CPU and GPU, the output results can be saved to this shared memory and read by subsequent CPU and GPU operations without additional copies between CPU and GPU buffers in main memory. In addition, unlike *coarse-grained* SVM, *fine-grained* SVM does not require explicit data mapping and unmapping operations since the hardware guarantees cache coherence for memory shared between CPU and GPU. Hence, this technique not only avoids copying data between CPU and GPU, but also eliminates the cost of data mapping operations.
2. To reduce the notification delay, we dispatch an active polling OpenCL kernel after each GPU computation. This kernel updates two synchronization variables (`cpu_flag` and `gpu_flag`) stored in fine-grained SVM. The kernel first updates `gpu_flag` to indicate GPU completion, then repeatedly checks `cpu_flag` to wait for CPU completion. Meanwhile, the CPU updates `cpu_flag` once finishing the computation and keeps polling for `gpu_flag` to be updated by the GPU. Notably, in order to minimize synchronization overhead, our polling-based implementation requires *busy waiting* on both CPU and GPU, which leads to additional power consumption in the case of unbalanced workload partitioning. However, our accurate latency predictions help balance CPU and GPU computation times and mitigate this issue.

Overall, by eliminating data mapping operations and reducing notification delay, we reduce the mean synchronization overhead to 7 μs on Motorola Edge Plus 2022 across 2,039 linear operations, significantly improving the efficiency of CPU-GPU co-execution.

## 5   Experimental Results

In this section, we conduct comprehensive evaluations of our co-execution strategy using 2,039 linear and 2,051 convolution operations on four mobile platforms: Pixel 4, Pixel 5, Motorola Edge Plus 2022 (Moto 2022), and OnePlus 11.

### 5.1   Experimental Setup

Building on TFLite, we developed a C++ benchmarking tool that co-executes OpenCL kernels from the TFLite GPU Delegate [12] and CPU kernels from the XNNPACK library [4]; CPU-GPU synchronization kernels were implemented in OpenCL and dispatched to the same GPU queue. The compiled binary was uploaded to each smartphone to benchmark the latency of co-execution.

To ensure stable performance measurements, we prepared the smartphones using the same configurations as [14]. Specifically, we enabled the performance mode on the GPU and all CPU cores to encourage near-maximum clock frequencies, reducing performance fluctuations due to dynamic frequency scaling on Android platforms. The CPU threads were scheduled to high-performance cores by specifying CPU affinity. In addition, we attached an external cooling fan to the back of the smartphones and allowed for a cool-down time (1 s) after profiling each operation configuration, in order to mitigate thermal throttling effects, which can significantly hinder sustained performance.

### 5.2   Training Dataset Generation and Latency Prediction Accuracy

We construct a training dataset for ML predictors by sampling a broad range of operation parameters. For linear layers, operation dimensions (input length $L$, input channels $C_{in}$, and output channels $C_{out}$) are selected using a structured random sampling approach: first, we randomly pick an interval from $\{[2^k, 2^{k+1}] \mid 2 \leq k \leq 9\}$, and then we sample the dimensions uniformly from the selected interval. For convolutional layers, we sample input height $H_{in}$, input width $W_{in}$, input channels $C_{in}$, and output channels $C_{out}$ using the same approach, and we sample the kernel (filter) shape $K$ from $\{1, 3, 5, 7\}$ and its stride $S$ from $\{1, 2\}$. In total, we collect latency measurements for 12,500 distinct configurations for each type of layer (linear or convolutional), using 20% for testing.

To train latency predictors, we use gradient-boosted decision trees (GBDTs) from LightGBM [10] and Optuna [1] to tune hyperparameters, which include learning rate (0.01 to 0.2), number of estimators (100 to 1000), depth (5 to 20), number of leaves (16 to 512), L1/L2 regularization terms ($10^{-8}$ to 1), and subsample ratios (0.5 to 1). As discussed in Sect. 3.2, we train separate predictors

**Table 1.** MAPEs of GBDT Predictors

| Device | Operations | MAPEs | | | |
|---|---|---|---|---|---|
| | | GPU | 1 CPU | 2 CPUs | 3 CPUs |
| Pixel 4 | Linear | 4.4% | 11.5% | 7.1% | 5.8% |
| | Convolutional | 8.5% | 11.4% | 8.8% | 7.2% |
| Pixel 5 | Linear | 3.7% | 6.2% | 7.8% | 7.2% |
| | Convolutional | 7.7% | 6.9% | 8.1% | 7.1% |
| Moto 2022 | Linear | 4.0% | 2.5% | 2.6% | 2.4% |
| | Convolutional | 9.0% | 4.0% | 3.6% | 3.5% |
| OnePlus 11 | Linear | 3.7% | 3.1% | 2.9% | 3.1% |
| | Convolutional | 7.4% | 4.8% | 4.2% | 4.4% |

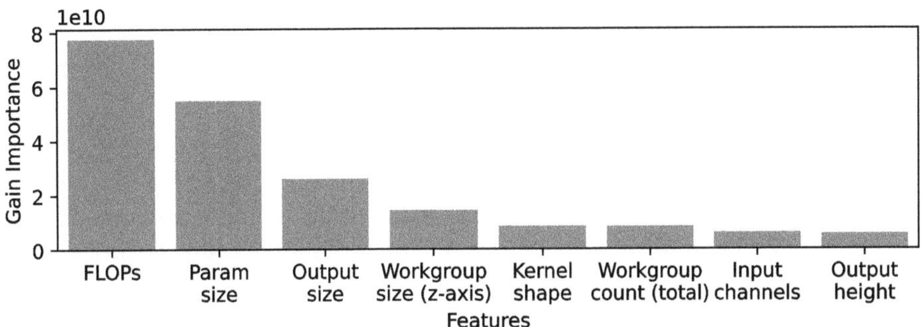


**Fig. 7.** Gain Improvement from GBDT Input Features (Convolution, Moto 2022)

for each kernel implementation, based on features including operation configurations and workgroup information. Figure 7 presents the gain importance (i.e., the total loss improvement for all splits of a feature) for the top eight features of convolutional layers. As shown, workgroup size and total workgroup count are important factors affecting latency, which motivates their inclusion as input features. The GBDT predictors typically take 3–4 ms to determine the optimal partitioning for each operation; these partitioning decisions can be made offline before deployment to a device, as part of the compilation process.

Table 1 presents the prediction Mean Average Percentage Error (MAPE); errors are generally higher for convolutions due to the greater number of parameters (e.g., filter shape, stride) and multiple kernel implementations.

## 5.3   Co-Execution Speedup of Individual Layers

Table 2 reports the speedups of our co-execution strategy, together with the best speedups found by a grid search over $[0, C_{out}]$ with step size equal to 8. Given the long times required for repeated measurements, grid search is evaluated only on a random subset including 10% of the test cases. Note that grid search is only a baseline that is not applicable to real-world systems due to the long

**Table 2.** Average Speedups Obtained through CPU-GPU Co-Execution

| Device | Method | Speedup of Linear | | | Speedup of Convolutional | | |
|---|---|---|---|---|---|---|---|
| | | 1 thread | 2 threads | 3 threads | 1 thread | 2 threads | 3 threads |
| Pixel 4 | GBDT | 1.21x | 1.52x | 1.84x | 1.22x | 1.46x | 1.69x |
| | Search | 1.29x | 1.59x | 1.92x | 1.31x | 1.56x | 1.79x |
| Pixel 5 | GBDT | 1.51x | 1.78x | 1.89x | 1.45x | 1.69x | 1.75x |
| | Search | 1.63x | 1.92x | 2.01x | 1.49x | 1.80x | 1.87x |
| Moto 2022 | GBDT | 1.20x | 1.32x | 1.44x | 1.16x | 1.27x | 1.39x |
| | Search | 1.23x | 1.36x | 1.49x | 1.22x | 1.34x | 1.46x |
| OnePlus 11 | GBDT | 1.06x | 1.17x | 1.26x | 1.07x | 1.22x | 1.35x |
| | Search | 1.13x | 1.25x | 1.35x | 1.12x | 1.27x | 1.40x |

measurement times required for each new set of parameters of the linear and convolution operations. Test cases are selected as follows.

- *Linear Layers:* Dimensions are selected from $\{i \cdot 2^j \mid 4 \leq i \leq 6, 2 \leq j \leq 9\}$. Then, we keep operations with FLOPs in $[4 \cdot 10^6, 10^9]$, resulting in a total of 2,039 linear operations. Our predictor leads to up to 1.89x average speedup (on Pixel 5), close to the best average speedup of 2.01x obtained through grid search.
- *Convolutional Layers:* Mobile vision models adopt hierarchical stages [16], with earlier stages capturing the local details (e.g., texture) and later stages capturing global features (e.g., shape). Accordingly, we define a hierarchy of 4 stages: convolutions of the first stage have input resolution $H_{in}, W_{in} \in \{64, 56, 48, 40\}$, kernel (filter) shape $K \in \{1, 3, 5, 7\}$, stride $S \in \{1, 2\}$, and input/output channels $C_{in}, C_{out} \in \{256/i, 320/i, 384/i, 448/i, 512/i\}$ where $i = 1, 1, 4, 8$ for $K = 1, 3, 5, 7$, respectively (to have similar computation loads). In stages 2, 3, 4, we halve the resolution and double the number of channels. We keep the generated layers with FLOPs within range $[4 \cdot 10^6, 10^9]$, resulting in a total of 2,051 convolutional layers. As shown in Table 2, our predictor achieves up to 1.75x speedup on Pixel 5, close to the best measured speedup of 1.87x. Notably, speedups are higher on devices with smaller CPU/GPU performance gap (e.g., Pixel 4 and Pixel 5), as a larger fraction of the task can be offloaded to the CPU.

## 5.4   Co-Execution Speedup of End-to-End Models

In this section, we present experiments demonstrating the speedups in end-to-end latency. We focus on four common neural networks: VGG16 [17], ResNet-18 [5], ResNet-34, and Inception-v3 [18]. In Table 3, we report the baseline latency for running each model on GPU. As comparison, we evaluate the performance of individual convolution and linear operations through co-execution on GPU and 3 CPU threads; we also conduct end-to-end experiments that incorporate the

**Table 3.** End-to-End Speedups from GPU and 3 CPU Threads Co-Execution

| Device | Neural Network | Baseline (ms) | Individual Ops | | End-to-End | |
|---|---|---|---|---|---|---|
| | | | Latency (ms) | Speedup | Latency (ms) | Speedup |
| Pixel 4 | VGG16 | 83.3 | 70.8 | 1.18x | 73.0 | 1.14x |
| | ResNet-18 | 17.5 | 11.0 | 1.59x | 11.4 | 1.54x |
| | ResNet-34 | 37.5 | 22.2 | 1.69x | 22.5 | 1.67x |
| | Inception-v3 | 99.5 | 59.1 | 1.68x | 61.5 | 1.62x |
| Pixel 5 | VGG16 | 194.8 | 119.8 | 1.63x | 125.1 | 1.56x |
| | ResNet-18 | 33.2 | 18.2 | 1.82x | 18.6 | 1.78x |
| | ResNet-34 | 65.2 | 36.9 | 1.77x | 37.1 | 1.76x |
| | Inception-v3 | 184.3 | 99.9 | 1.85x | 102.9 | 1.79x |
| Moto 2022 | VGG16 | 32.0 | 28.7 | 1.11x | 29.7 | 1.08x |
| | ResNet-18 | 7.5 | 6.3 | 1.18x | 6.7 | 1.11x |
| | ResNet-34 | 14.7 | 12.2 | 1.20x | 12.9 | 1.14x |
| | Inception-v3 | 52.0 | 38.8 | 1.34x | 41.1 | 1.27x |
| OnePlus 11 | VGG16 | 27.4 | 25.3 | 1.09x | 26.2 | 1.05x |
| | ResNet-18 | 8.5 | 6.6 | 1.29x | 6.8 | 1.25x |
| | ResNet-34 | 17.6 | 13.5 | 1.31x | 13.8 | 1.27x |
| | Inception-v3 | 44.2 | 36.2 | 1.22x | 37.8 | 1.17x |

offline partitioning decisions for each operation and correspondingly schedule CPU and GPU kernels for the entire model; notably, pooling operations are always scheduled on the GPU, since their latency is negligible and this can avoid the synchronization overhead. The end-to-end improvement is slightly lower than that of individual operations, potentially due to memory access overhead between layers. The results show that our proposed method can achieve up to 1.67x, 1.79x, 1.27x, and 1.27x average speedups on Pixel 4, Pixel 5, Moto 2022, and OnePlus 11, respectively; these results validate the effectiveness of approach to real-world neural networks.

We observe that related work [9] also evaluated co-execution of VGG16, reducing its inference latency on Pixel 4 from a baseline of 200 ms (using only the GPU) to around 150 ms (using CPU and GPU). In contrast, our evaluation of VGG16 on Pixel 4 started from a baseline of 83.3 ms, which was reduced to 73.0 ms through co-execution. The difference in baseline inference latency is due to the different performance of the MACE ML framework [21] (used in [9]) and TFLite (used in our work). In particular, TFLite (1) already leverages image storage types to take advantage of L1 texture cache (which was used in [9] to improve the MACE baseline) and (2) implements efficient Winograd kernels to accelerate convolutional layers in VGG16 (Sect. 3.2).

## 5.5 Ablation Study

We conduct an ablation study on Moto 2022 to evaluate the individual impact of our proposed techniques. First, we observe that our augmentation technique

**Table 4.** Ablation Study: Co-Execution Speedup (Moto 2022)

| Method | Speedup of Linear | | | Speedup of Convolutional | | |
|---|---|---|---|---|---|---|
| | 1 thread | 2 threads | 3 threads | 1 thread | 2 threads | 3 threads |
| Ours | 1.20x | 1.32x | 1.44x | 1.16x | 1.27x | 1.39x |
| w/o Augmentation | 1.12x | 1.24x | 1.37x | 1.08x | 1.19x | 1.31x |
| Original Overhead | 0.76x | 0.81x | 0.88x | 0.98x | 1.07x | 1.17x |

reduces latency prediction MAPE of linear layers from 9.3% to 4.4% and of convolutional layers from 14.1% to 9.3%. These improvements lead to better partitioning strategies, e.g., for convolutional layers using the GPU and 1 CPU thread, the latency reduction improves from 1.08x to 1.16x (Table 4).

Next, to evaluate the contribution of our overhead reduction technique, we compare our active polling implementation with a baseline where the CPU *passively* waits for GPU kernel completions using the OpenCL `clWaitForEvents` API. The baseline incurs average overhead of 162 µs across 2,039 linear layers and 141 µs across 2,051 convolutional layers, accounting for 39.9% and 15.8% of the total co-execution latency, respectively, in the case of GPU co-execution with 1 CPU thread. Such high overhead negates the benefits of co-execution; in contrast, our active polling approach makes the synchronization overhead negligible (average of 7.0 µs for linear layers and 5.4 µs for convolutional layers).

## 6   Conclusions

We explored inference latency optimization for deep neural networks on mobile platforms by partitioning individual linear and convolutional layers across CPU and GPU. To address challenges in accurately predicting complex GPU latency behaviors and reducing CPU-GPU synchronization overhead, we developed enhanced latency predictors incorporating kernel implementation and dispatching information, and we proposed a lightweight synchronization method using OpenCL shared virtual memory. Comprehensive experimental evaluations showed that our approach achieves significant speedups that are close to the achievable best. In future work, we plan to investigate parallel execution on CPU, GPU, and NPU, and the effects of model quantization.

**Acknowledgments.** This work was supported in part by the NSF CNS-1816887, CCF-1763747, and IIS-1833137 awards. The authors would like to thank the anonymous EPEW reviewers for their insightful comments that helped improve this paper.

# References


1. Akiba et al.: Optuna: a next-generation hyperparameter optimization framework. In: Proceedings of KDD, pp. 2623–2631 (2019)
2. Chen et al.: HeteroLLM: accelerating large language model inference on mobile SoCs platform with heterogeneous AI accelerators. arXiv:2501.14794 (2025)
3. Dosovitskiy et al.: An image is worth $16 \times 16$ words: transformers for image recognition at scale. arXiv:2010.11929 (2020)
4. Google: XNNPACK: high-efficiency floating-point neural network inference operators for mobile, server, and web. https://github.com/google/XNNPACK (2023)
5. He, K., Zhang, X., Ren, S., Sun, J.: Deep residual learning for image recognition. In: Proceedings of CVPR, pp. 770–778. IEEE (2016)
6. Intel: OpenVINO Toolkit (2025). https://github.com/openvinotoolkit
7. Jang et al.: Sparsity-aware and re-configurable NPU architecture for Samsung flagship mobile SoC. In: Proceedings of ISCA, pp. 15–28. IEEE (2021)
8. Jayanth, R., Gupta, N., Prasanna, V.: Benchmarking edge AI platforms for high-performance ML inference. arXiv:2409.14803 (2024)
9. Jia et al.: CoDL: efficient CPU-GPU co-execution for deep learning inference on mobile devices. In: Proceedings of MobiSys, vol. 22, pp. 209–221 (2022)
10. Ke et al.: LightGBM: a highly efficient gradient boosting decision tree. Proc. NeurIPS **30** (2017)
11. Kim, Y., Kim, J., Chae, D., Kim, D., Kim, J.: $\mu$layer: low latency on-device inference using cooperative single-layer acceleration and processor-friendly quantization. In: Proceedings of EuroSys, pp. 1–15 (2019)
12. Lee et al.: On-device neural net inference with mobile GPUs. arXiv:1907.01989 (2019)
13. Li, Z., Paolieri, M., Golubchik, L.: Predicting inference latency of neural architectures on mobile devices. In: Proceedings of ICPE, pp. 99–112. ACM (2023)
14. Li, Z., Paolieri, M., Golubchik, L.: A benchmark for ML inference latency on mobile devices. In: Proceedings of EdgeSys, pp. 31–36. ACM (2024)
15. Li, Z., Paolieri, M., Golubchik, L.: Inference latency prediction for CNNs on heterogeneous mobile devices and ML frameworks. Perform. Eval. **165**, 102429 (2024)
16. Sandler, M., Howard, A., Zhu, M., Zhmoginov, A., Chen, L.C.: MobileNetV2: inverted residuals and linear bottlenecks. In: Proceedings of CVPR, pp. 4510–4520 (2018)
17. Simonyan, K., Zisserman, A.: Very deep convolutional networks for large-scale image recognition. arXiv preprint arXiv:1409.1556 (2014)
18. Szegedy, C., Vanhoucke, V., Ioffe, S., Shlens, J., Wojna, Z.: Rethinking the inception architecture for computer vision. In: Proceedings of CVPR, pp. 2818–2826. IEEE (2016)
19. Tang et al.: To bridge neural network design and real-world performance: a behaviour study for neural networks. Proc. MLSys **3**, 21–37 (2021)
20. Wei, J., et al.: NN-Stretch: automatic neural network branching for parallel inference on heterogeneous multi-processors. In: Proceedings of MobiSys, pp. 70–83 (2023)
21. XiaoMi: Mace: a deep learning inference framework for mobile heterogeneous computing platforms (2025). https://github.com/XiaoMi/mace. Accessed 27 Aug 2025
22. Zhang et al.: NN-Meter: towards accurate latency prediction of deep-learning model inference on diverse edge devices. In: Proceedings of MobiSys, pp. 81–93 (2021)

# Approximating Heavy-Tailed Distributions with a Mixture of Bernstein Phase-Type and Hyperexponential Models


Abdelhakim Ziani[1]([✉]), András Horváth[2], and Paolo Ballarini[1]

[1] Université Paris Saclay, Lab. MICS, CentraleSupélec, Gif-sur-Yvette, France
{hakim.ziani,paolo.ballarini}@centralesupelec.fr
[2] Università di Torino, Turin, Italy
horvath@di.unito.it



**Abstract.** Heavy-tailed distributions, prevalent in a lot of real-world applications such as finance, telecommunications, queuing theory, and natural language processing, are challenging to model accurately owing to their slow tail decay. Bernstein phase-type (BPH) distributions, through their analytical tractability and good approximations in the non-tail region, can present a good solution, but they suffer from an inability to reproduce these heavy-tailed behaviors exactly, thus leading to inadequate performance in important tail areas. On the contrary, while highly adaptable to heavy-tailed distributions, hyperexponential (HE) models struggle in the body part of the distribution. Additionally, they are highly sensitive to initial parameter selection, significantly affecting their precision.

To solve these issues, we propose a novel hybrid model of BPH and HE distributions, borrowing the most desirable features from each for enhanced approximation quality. Specifically, we leverage an optimization to set initial parameters for the HE component, significantly enhancing its robustness and reducing the possibility that the associated procedure results in an invalid HE model. Experimental validation demonstrates that the novel hybrid approach is more performant than individual application of BPH or HE models. More precisely, it can capture both the body and the tail of heavy-tailed distributions, with a considerable enhancement in matching parameters such as mean and coefficient of variation. Additional validation through experiments utilizing queuing theory proves the practical usefulness, accuracy, and precision of our hybrid approach.


## 1 Introduction

Heavy-tailed distributions are common in real-world domains like finance, telecommunications, queuing theory, and NLP, where rare or extreme events significantly influence system behavior [1]. Markov models, while advantageous due to their analytical and numerical tractability [2], struggle to represent these distributions accurately in the tail region.

Bernstein Phase-Type (BPH) distributions [3,4], a class of distributions defined in terms of Markov chains, offer fast parameter estimation and precise approximation of the body of a distribution, but fall short in capturing heavy tails. Hyperexponential (HE) distributions, on the other hand, can better approximate slowly decaying tails when used with proper parameters [5] but (i) they are highly sensitive to initialization, which can hinder performance, and (ii) they form an inflexible family of distributions (their probability density function can only be monotone decreasing).

To address these limitations, we propose a hybrid model combining BPH and HE distributions. This approach leverages the strengths of both: the approximation efficiency of BPH distributions and the tail-fitting flexibility of HE distributions. To improve the robustness of the HE component, we employ the Adam optimizer to enhance parameter initialization. This paper presents the hybrid BPH_HE model and the optimization method, evaluates performance, and validates results through queuing theory applications.

A similar approach was presented in [6], where the authors proposed combining acyclic phase-type (APH) distributions with HE distributions. The improvements over that work are as follows: (i) instead of APH, we apply BPH distributions, which allow for much faster and more precise approximation of the body, and (ii) we automate the parameter initialization of the HE distributions that capture the tail, whereas in [5] and [6], this process is left to a non-trivial manual setting.

The remainder of the paper is organized as follows: Sect. 2 reviews related work, Sect. 3 gives background, Sect. 4 details the hybrid model and the optimization strategy, Sect. 5 covers experiments, Sect. 6 application results, and Sect. 7 concludes the study.

## 2 Related Work

Phase-type (PH) distributions have been introduced in [7] to model non-exponential durations by combination of exponential distributions, maintaining thus the Markov property of the underlying stochastic process. The application of PH distributions requires fitting (also called matching) algorithms, that is, methods that, given a distribution, provide a PH distribution that closely resembles it. Such methods either aim

1. to match statistical properties extracted from the distribution (typically moments) or
2. to minimize a distance measure that considers the whole distribution.

A seminal work among those belonging to the first category is [8], where the authors use a mixture of Erlang distributions to match the first three moments of a given distribution. This approach was refined in [9], providing a method to match any valid first three moments with an acyclic PH distribution of minimal size. Matching of more than three moments was tackled in [10]. While the above approaches match the moments exactly, there have been proposals to approximate matching of the moments based on optimization [11,12]. A common difficulty of all these approaches is that a full characterization of the feasible region of moments of PH distributions is not available. The known results in this direction are gathered in [13].

Approaches belonging to the second category are most often based on the maximum likelihood principle, that is, they minimize the KullbackLeibler divergence. This can be done either by using the Expectation-Maximization (EM) algorithm, like in [14,15], or by more direct maximization of the likelihood function, as done in [16], where also a benchmark for the evaluation of fitting methods is proposed.

For what concerns fitting heavy tailed distributions, general approaches belonging to either the first or the second category fail to result in PH distributions with satisfactory behavior. For this reason, [5] developed a heuristic approach, which we describe in detail in Sect. 3.4, whose shortcoming is that it is applicable only to distributions with a monotone probability density function. To overcome this, in [6] the authors proposed to combine the model proposed in [5] with an acyclic PH distribution whose parameters are estimated based on the maximum likelihood principle.

Models constructed through the composition of PH distributions (see, e.g., [17]) result in Markov chains, which can be analyzed using established techniques developed for such models [2]. In the area of queuing systems, a wide range of methods have been developed to analyze models that employ PH distributions. The seminal work in this regard is [18] while [19] provides a modern, comprehensive treatment. A first effort to propose a tool to analyze Petri net models with PH timing was described in [20].

Among others, tools for PH fitting are [21,22] while [23] provides both moment matching algorithms and methods to analyze queues with PH timing.

## 3   Background

**Notations.** Given a positive continuous random variable, $X$, we denote by $f(x)$ its probability distribution function (PDF), $F(x) = \int_0^x f(x)dx$ its cumulative distribution function (CDF) and $\bar{F}(x) = 1 - F(x)$ its complementary cumulative distribution function (CCDF).

### 3.1   Heavy Tailed Distributions

A heavy-tailed distribution is a probability distribution whose tail decays more slowly than that of an exponential distribution, implying that very large values can occur with non-negligible probability [1]. Mathematically, a distribution $F(x)$ is said to be heavy-tailed if its CCDF satisfies

$$\lim_{x \to \infty} e^{\lambda x} \bar{F}(x) = \infty, \quad \forall \lambda > 0 \tag{1}$$

meaning that the probability of large values does not decay at least exponentially fast, leading to distributions with a significant probability mass in the tail region. In most practical applications, heavy-tailed distributions are defined on positive values.

Heavy-tailed distributions appear in a wide range of real-world phenomena, often characterizing situations where extreme values or rare events have a non-negligible impact. Examples include queuing theory [24,25], where in many fields such as networking, telecommunication, or cloud computing, service time is modeled by a heavy-tailed distribution. In finance, risk modeling frequently exhibits heavy tails [26], meaning extreme outcomes or losses occur more often than in a Gaussian model (where rare events happen with a probability near zero), which is crucial in portfolio management. In natural language processing (NLP), word frequency distributions in all spoken languages follow a heavy-tailed distribution, characterized by Zipf's law [27], where 20% of the most frequent words make 80% of the spoken or written language. This property is particularly important in large language models (LLMs), as they must balance learning common words and patterns while still capturing rare but meaningful linguistic structures [28].

## 3.2   Phase Type Distributions

A phase-type (PH) distribution is a probability distribution resulting from the combination of exponential distributions. More formally, as introduced in [7], a degree $n$ continues PH distribution is given by the time to absorption in a continuous-time Markov chain (CTMC) having $n$ transient states (called phases) and one absorbing state. The most widely used representation of a PH distribution is a vector-matrix pair $(a, A)$ where the initial probability vector $a$ and the infinitesimal generator matrix $A$ only account for the transient states of the Markov chain[1]. A graphical example of a PH distribution is provided in Fig. 1 with vector-matrix pair representation

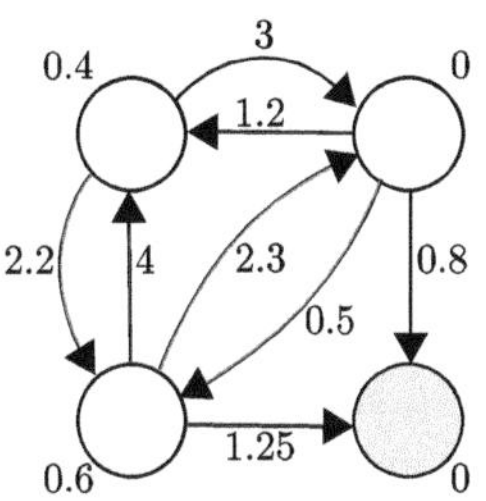


**Fig. 1.** A degree 3 PH distribution

$$a = \begin{bmatrix} 0.4\ 0\ 0.6 \end{bmatrix}, \quad A = \begin{bmatrix} -5.2 & 3 & 2.2 \\ 1.2 & -2.5 & 0.5 \\ 4 & 2.3 & -7.55 \end{bmatrix}.$$

Given $a$ and $A$ of a PH distribution, the associated PDF, CDF, and CCDF are given by

$$f(x) = ae^{xA}(-A\mathbf{1}), \ F(x) = 1 - ae^{xA}\mathbf{1}, \ \text{and} \ \bar{F}(x) = ae^{xA}\mathbf{1} \tag{2}$$

where $\mathbf{1}$ denotes a column vector in which all entries are equal to 1.

---

[1]  This is because the intensities from the transient states to the absorbing one can be determined based on $A$ and we assume that the initial probability associated with the absorbing state is 0.

Phase-type distributions form a dense family of distributions, meaning that any distribution can be approximated arbitrarily well by a PH distribution. However, since PH distributions are inherently light-tailed (i.e., their tails decay exponentially), classical parameter estimation techniques, such as maximum likelihood estimation or moment matching, typically fail to approximate heavy-tailed distributions accurately. To address this, PH distributions with special structures must be employed. From an application standpoint, PH distributions offer both analytical and numerical tractability, providing closed-form expressions for key metrics such as moments, tail probabilities, and Laplace transforms.

### 3.3   Bernstein Phase Type Distributions

Given a function $g : [0, 1] \rightarrow \mathbb{R}$, its order-$n$ Bernstein polynomial (BP) approximation is given by

$$BP_n(g; x) := \sum_{i=0}^{n} g\left(\frac{i}{n}\right) \cdot \binom{n}{i} x^i (1 - x)^{n-i} \,, \tag{3}$$

which is a degree-$n$ polynomial.

With the change of variable $y = e^{-x}$ (i.e., $x = -\log(y)$), which maps the interval $[0, 1]$ onto $[0, \infty)$, one can approximate a function $f : [0, \infty) \rightarrow \mathbb{R}$. This leads to the Bernstein exponential (BE) approximations in the form of

$$BE_n(f; x) := \sum_{i=0}^{n} f\left(\log\left(\frac{n}{i}\right)\right) \cdot \binom{n}{i} e^{-ix} (1 - e^{-x})^{n-i} \tag{4}$$

where the division by zero in case of $i = 0$ is resolved by considering the limiting value of $f(x)$ as $x$ tends to infinity, that is, $f\left(\log \frac{n}{0}\right) = \lim_{x \to \infty} f(x)$. The approximation given in Eq. (4) can be applied either to the PDF, to the CDF or to the CCDF of a probability distribution. In all cases, the approximation belongs to a class of PH distributions that we call Bernstein phase-type (BPH) distributions [4]. In this paper, we apply CDF based approximation as it is somewhat more straightforward than using the PDF [4]. Figure 2 provides a graphical representation of the obtained PH distribution when Eq. (4) is applied to the CDF $F(x)$. The vector of initial probabilities

$$a = [F(\infty) - F(\log n) \quad F(\log n) - F(\log(n/2)) \quad ... \quad F(\log(n/(n-1))) - F(0)]$$

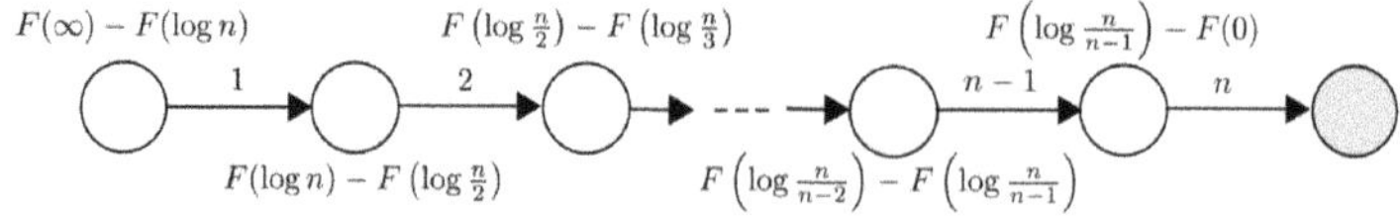


**Fig. 2.** Bernstein PH approximation of a CDF $F(x)$.

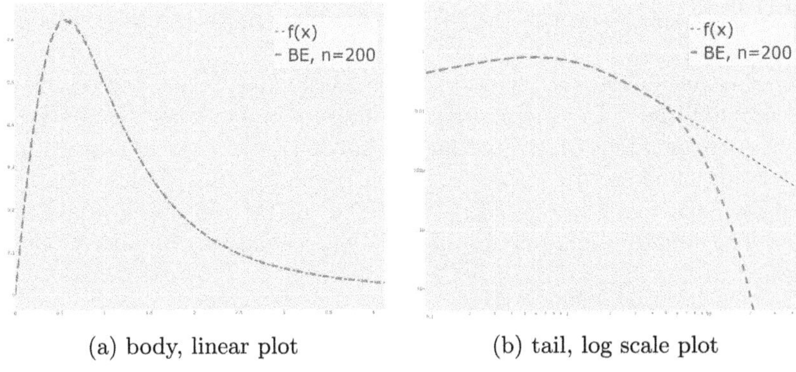


(a) body, linear plot                    (b) tail, log scale plot

**Fig. 3.** PDF resulting from a CDF-based BPH approximation of the Burr distribution with parameters $c = 2, d = 1$.

are calculated directly based on the CDF while the intensities are fixed, in the sense that they do not depend on the CDF we approximate. If the CDF corresponds to a normalized distribution without mass at 0, then $F(\infty) = 1$ and $F(0) = 0$. In [4] it was shown that applying Eq. (4) to a CDF $F(x)$ or to its corresponding CCDF $\bar{F}(x) = 1 - F(x)$ results in the very same BPH distribution. The initial probabilities based on the CCDF are simply

$$a = \left[\bar{F}(\log n) - \bar{F}(\infty) \quad \bar{F}(\log(n/2)) - \bar{F}(\log n) \quad ... \quad \bar{F}(0) - \bar{F}(\log(n/(n-1)))\right]$$

Figure 3 shows the PDF resulting from a CDF-based BPH approximation[2] of the Burr distribution, whose PDF is

$$f(x) = c \cdot d \cdot \frac{x^{c-1}}{(1 + x^c)^{d+1}} \, ,$$

using parameters $c = 2, d = 1$. A good approximation can be obtained in the body part of the distribution, especially when using a considerably large $n$ value. However, BPH models result in poor approximation in the tail region due to the fact that they lack the flexibility to capture the slow decay of heavy-tailed distributions.

## 3.4   Hyperexponential Approximation of Heavy-Tailed Distributions

In [5] a method is proposed to approximate heavy-tailed distributions with a mixture of exponential distributions whose CCDF is in the form:

$$\bar{F}(x) = \sum_{i=1}^{k} p_i e^{-\lambda_i x} \quad \text{with} \quad \sum_{i=1}^{k} p_i = 1 \quad \text{and} \quad 0 \le p_i \le 1, \lambda_i > 0, i = 1, 2, ..., k \quad (5)$$

---

[2] We prefer to show the PDF because it highlights more the differences than plotting the CDF or the CCDF.

which is also a special case of the PH distributions, called the hyperexponential (HE) distribution.

The mixture in Eq. (5) has $2k - 1$ free parameters ($k$ intensities and $k - 1$ initial probabilities). The method aims to find parameters for the mixture such that the corresponding CCDF is close to the CCDF we need to approximate at $2k - 1$ points, denoted by $x_1, x_2, ..., x_{2k-1}$. Assuming that we have $x_1 > x_2 > ... > x_{2k-1}$ and denoting by $\bar{F}(x)$ the CCDF to fit, the method proceeds as follows. The first two parameters, $p_1$ and $\lambda_1$, are chosen in such a way that we have

$$p_1 e^{-\lambda_1 x_1} = \bar{F}(x_1) \quad \text{and} \quad p_1 e^{-\lambda_1 x_2} = \bar{F}(x_2) \tag{6}$$

The further pairs, $p_i$ and $\lambda_i$ with $2 \leq i \leq k - 1$, are determined based on

$$p_i e^{-\lambda_i x_{2i-1}} = \bar{F}(x_{2i-1}) - \sum_{j=1}^{i-1} p_j e^{-\lambda_j x_{2i-1}} \quad \text{and} \quad p_i e^{-\lambda_i x_{2i}} = \bar{F}(x_{2i}) - \sum_{j=1}^{i-1} p_j e^{-\lambda_j x_{2i}} \tag{7}$$

That is, the effect of the already determined terms is taken into account by decreasing $\bar{F}(x)$. Finally, $p_k$ is computed as

$$p_k = 1 - \sum_{i=1}^{k-1} p_k$$

and $\lambda_k$ based on the last point $x_{2k-1}$ in such a way that

$$p_k e^{-\lambda_k x_{2k-1}} = \bar{F}(x_{2k-1}) - \sum_{j=1}^{k-1} p_j e^{-\lambda_j x_{2k-1}}$$

All the above calculations correspond to simple explicit expressions to determine the parameters. The basic idea behind the procedure is that the term $p_i e^{-\lambda_i x}$, whose parameters are determined based on $\bar{F}(x_{2i-1})$ and $\bar{F}(x_{2i})$, have a negligible effect at the points $x_j$ with $j > 2i$ because of the exponential decay of the term itself.

If the points $x_i$ are not chosen appropriately, the calculations proposed in [5] may lead to a non-valid hyperexponential distribution (e.g., with negative initial probabilities) or poor approximation. Since [5] does not provide an algorithm for selecting these points, applying the method remains challenging.

Figure 4 shows the HE approximation of the Burr distribution with $c = 2, d = 1$ using the method in [5]. While the approximation captures the tail behavior well over a long interval, the figures reveal an important weakness of HE approximations: they come with a monotone PDF and hence cannot properly approximate the body region where the Burr PDF is non-monotonic.

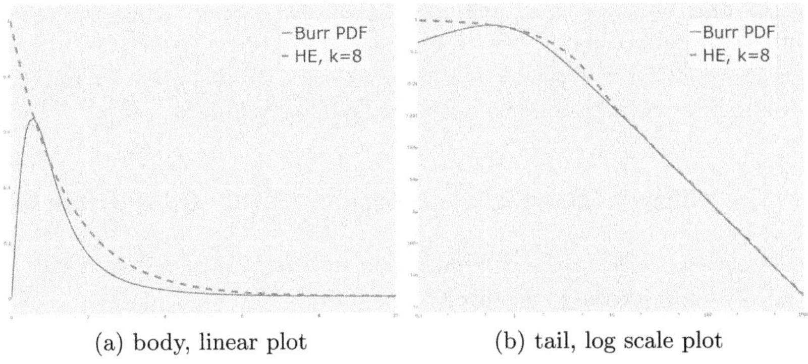

(a) body, linear plot                    (b) tail, log scale plot

**Fig. 4.** PDF of the HE approximation of a Burr distribution


# 4    Methodology

## 4.1    Approximation with Mixture of BPH and HE Distributions

In order to overcome the limitations of the previously described models, we combine the HE approximation (Sect. 3.4) and the BPH approximation (Sect. 3.3), exploiting the strength of the two methods. The resulting mixture, called BPH_HE distribution, belongs to the class of PH distributions.

For what concerns the HE component that approximates the tail, the procedure is almost the same as the one described in Sect. 3.4. The only difference is that the sum of the initial probabilities associated with the exponential terms has to be less than 1, resulting in a defective HE distribution, which will be combined with an order $n$ BPH distribution to obtain the final BPH_HE distribution. Accordingly, having one less constraint, the parameters $p_i, \lambda_i, i = 1, ..., k$ are determined based on $2k$ points instead of $2k-1$, and, instead of $\sum_{i=1}^{k} p_i = 1$, we must have $\sum_{i=1}^{k} p_i < 1$. As before, explicit expressions are used to calculate $p_i, \lambda_i, i = 1, ..., k$ that can be derived easily based on Eq. (6) and Eq. (7).

In order to combine the mixture of $k$ exponentials with a BPH, we introduce

$$\bar{G}(x) = \bar{F}(x) - \sum_{i=1}^{k} p_i e^{-\lambda_i x} \tag{8}$$

which is the CCDF $\bar{F}$ we aim to approximate minus the already determined contribution of the $k$ exponential terms. To obtain the BPH component of the mixture, we apply the approximation given in Eq. (4) to $\bar{G}$. Overall, the mixture of the $k$ exponential terms (the HE component) and the order $n$ BPH component has the CCDF

$$\hat{\bar{F}}(x) = \sum_{i=1}^{n} \bar{G}\left(\log \frac{n}{i}\right) \binom{n}{i} e^{-ix}(1 - e^{-x})^{n-i} + \sum_{j=1}^{k} p_j e^{-\lambda_j x} \tag{9}$$

where the first term is the contribution of the BPH part[3] while the second is that of the HE component.

The above CCDF corresponds to an order $n+k$ PH distribution. In its $(a, A)$ vector-matrix representation the initial probability vector is $a = \begin{bmatrix} a^{(\text{BPH})} & a^{(\text{HE})} \end{bmatrix}$ where

$$a^{(\text{BPH})} = \begin{bmatrix} \bar{G}(\log n) & \bar{G}(\log(n/2)) - \bar{G}(\log n) & \ldots & \bar{G}(0) - \bar{G}(\log(n/(n-1))) \end{bmatrix}$$

and $a^{(\text{HE})} = \begin{bmatrix} p_1 & p_2 & \ldots & p_k \end{bmatrix}$. Normalization of $a$ is guaranteed by $\bar{G}(0) = 1 - \sum_{j=i}^{k} p_i$. The infinitesimal generator is

$$A = \begin{bmatrix} A^{(\text{BPH})} & \mathbf{0} \\ \mathbf{0} & A^{(\text{HE})} \end{bmatrix}$$

where $\mathbf{0}$ denotes zero matrices of appropriate size, indicating that there are no transitions from states of the HE component to states of the BPH component and vice versa. $A^{(\text{BPH})} \in \mathbb{R}^{n \times n}$ is a bidiagonal matrix corresponding to the BPH component, and $A^{(\text{HE})} \in \mathbb{R}^{k \times k}$ is a diagonal matrix for the HE component. These are defined respectively as

$$A^{(\text{BPH})} = \begin{bmatrix} -1 & 1 & 0 & \cdots & & 0 \\ 0 & -2 & 2 & \cdots & & 0 \\ \vdots & \ddots & \ddots & & \ddots & \vdots \\ 0 & \cdots & 0 & -(n-1) & n-1 \\ 0 & \cdots & \cdots & & 0 & -n \end{bmatrix} \qquad A^{(\text{HE})} = \begin{bmatrix} -\lambda_1 & 0 & \cdots & & 0 \\ 0 & -\lambda_2 & \cdots & & 0 \\ \vdots & \vdots & \ddots & & \vdots \\ 0 & 0 & -\lambda_{k-1} & 0 \\ 0 & 0 & \cdots & & -\lambda_k \end{bmatrix}$$

## 4.2   Optimizing Data Initialization for the HE Tail Approximation

As already mentioned, the method proposed in [5] requires a set of points where the HE CCDF aims to closely match the target CCDF. The selection of these points is a non-trivial task.

The main objective of this part of the study is to find the optimal set of points, which we call below parameters, to initialize the HE tail approximation of the BPH_HE model. To achieve this, we formulate a supervised regression problem where the target values are the actual probabilities given by the CCDF we aim to approximate (denoted by $\bar{F}$), and the predictions are provided by the CCDF of the approximating BPH_HE (denoted by $\hat{\bar{F}}$).

The two CCDFs will be compared at a number of points $z_1, z_2, \ldots, z_n$, and, following the standard notation in supervised learning, we denote the value of $\bar{F}$ and $\hat{\bar{F}}$ at these points by:

$$y_i = \bar{F}(z_i) \quad \text{and} \quad \hat{y}_i = \hat{\bar{F}}(z_i)$$

---

[3] With respect to Eq. (4) the term with $i = 0$ is not present since we approximate the function $\bar{G}$ which tends to 0.

A good approximation will generate similar $y_i$ and $\hat{y}_i$ values. The points $z_1, ..., z_n$ can be selected to reflect the region of interest of the target CCDF. For simplicity, $z_i$ are chosen to be $n$ evenly spaced points over the domain of interest. While focusing on the tail could improve precision in that region, uniform spacing ensures better stability and convergence during optimization with gradient-based methods like Adam [29].

Two of the most applied loss functions in regression are the Mean Squared Error (MSE) and the Mean Absolute Error (MAE):

$$\text{MSE} = \frac{1}{n} \sum_{i=1}^{n} (y_i - \hat{y}_i)^2, \quad \text{MAE} = \frac{1}{n} \sum_{i=1}^{n} |y_i - \hat{y}_i|$$

Given the small differences in the tail region, using the MAE in this context is a more convenient choice.

Due to the fact that the model is probabilistic, it has a set of constraints to follow, namely, we must have $0 < p_i < 1$ and $\lambda_i > 0$. To ensure that the optimizer does not converge to solutions that violate these constraints, the original MAE loss is penalized when such constraints are not respected. The final loss function consists of three components: the Mean Absolute Error, a penalty term that enforces the constraint $0 < p_i < 1$, and an additional penalty that prevents negative values of $\lambda_i$.

Mathematically, the loss $\mathcal{L}$ is formulated as follows:

$$\mathcal{L} = w_{\text{mae}} \cdot \frac{1}{n} \sum_{i=1}^{n} |y_i - \hat{y}_i| + w_\lambda \cdot \frac{1}{k} \sum_{i=1}^{k} \mathbf{1}(\lambda_i < 0)$$
$$+ w_p \cdot \sum_{i=1}^{k} [\max(0, -p_i) + \max(0, p_i - 1)] \tag{10}$$

where $w_{\text{mae}}$, $w_\lambda$ and $w_p$ are the normalized weights for each loss term, $\mathbf{1}(\lambda_i \leq 0)$ is an indicator function to count the number of negative values in $\lambda$ and $\sum_{i=1}^{k} [\max(0, -p_i) + \max(0, p_i - 1)]$ penalizes $p_i$ values that fall outside the range $[0, 1]$.

The above loss function has been optimized using an Adam optimizer, a gradient descent-like algorithm, with initialization values following a power-of-two law: $x_i = 2^{i-1}$ with $i = 1, 2, ..., 2k$ where $k$ is the number of exponential components. With this initialization, the tail is covered in the interval $[1, 2^{k-1}]$. If the tail region we aim to consider is different and it is given by $[x_{\min}, x_{\max}]$, then we use the following initialization parameters

$$x_i' = x_{\min} + \frac{(x_i - x_1)(x_{\max} - x_{\min})}{x_{2k} - x_1} \quad i = 1, 2, ..., 2k$$

which maps the interval $[1, 2^{k-1}]$ to the interval $[x_{\min}, x_{\max}]$ through a linear transformation.

## 5   Experimental Setup

To assess the approximation accuracy, we run a number of experiments on well-known heavy-tailed distributions. Figure 5 shows the approximation for the previously used Burr distribution and a Pareto-type distribution with CCDF

$$\bar{F}(x) = \frac{1}{(x+1)^{3.1}}$$

with the corresponding BPE_HE approximations. Plots highlight a very good approximation of both the body and tail region of the distributions.

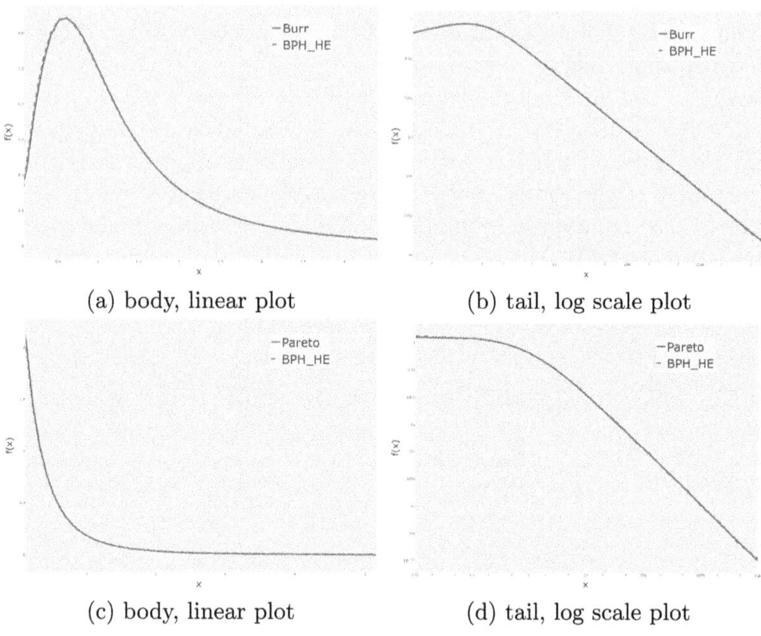

(a) body, linear plot          (b) tail, log scale plot

(c) body, linear plot          (d) tail, log scale plot

**Fig. 5.** PDF of BPH_HE approximation of a Burr distribution (a, b) and a Pareto distribution (c, d). The Burr case corresponds to Figs. 3 and 4.

One significant challenge in fitting probability distributions to real-world data is accurately matching statistical indices, particularly the mean and the coefficient of variation. Heavy-tailed distributions are particularly challenging because their tail behavior greatly influences these indices. If the fitted distribution does not closely match the target probability distribution, using it in real-world applications can lead to imprecise conclusions.

Table 1 compares the considered models, BPH, HE, and BPH_HE, from the perspective of capturing the mean and the coefficient of variation (CV) of a Pareto distribution. The combined model, that is, BPH_HE distributions, clearly outperforms BPH and HE distributions.

**Table 1.** Statistical indices across various models on a Pareto distribution.

| Model | MAE | Mean (Real) | Mean (Approx) | Coef of Var (Real) | Coef of Var (Approx) |
|---|---|---|---|---|---|
| BPH | 0.000019 | 0.52877 | 0.525738 | 1.444789 | 1.170760 |
| HE | 0.000014 | 0.52877 | 0.541833 | 1.444789 | 1.416929 |
| BPH_HE | **0.000001** | 0.52877 | **0.528773** | 1.444789 | **1.449654** |

Table 2 presents results obtained from fitting the BPH_HE model to previously described Burr and Pareto, and other heavy-tailed cases such as the Weibull and Lognormal, whose CCDF are given by, respectively:

$$\bar{F}(x) = e^{-\left(\frac{x}{5}\right)^{0.2}}$$

and

$$\bar{F}(x) = 1 - \Phi\left(\frac{\ln(x) - 1}{2}\right)$$

respectively, where $\phi(\cdot)$ denotes the CDF of the standard normal distribution.

The Mean Absolute Error remains extremely small, showing that the approximation is highly accurate. In case of the Pareto distribution, the mean and the coefficient of variation are reproduced with near-perfect precision, while for more complex cases like the Lognormal or Weibull distributions, the approximations remain remarkably close. These results demonstrate the robustness and flexibility of the BPH_HE approach, ensuring accurate modeling of both the body and the tail of heavy-tailed distributions.

## 6   Queuing System Experiment

To validate our method for approximating heavy-tailed distributions using the BPH_HE model, we apply it in a queuing scenario. Many queuing systems, such as in network analysis, exhibit heavy-tailed service times due to occasional large tasks that cause high variability. These distributions are mathematically complex, making direct analysis difficult. By replacing the original heavy-tailed service time in an M/G/1 queue with its PH approximation (yielding an M/PH/1 queue), we obtain a tractable alternative.

**Table 2.** Comparison of statistical metrics for different distributions.

| Distribution | MAE | Mean (Real) | Mean (Approx) | Coef. of Var (Real) | Coef of Var (Approx) |
|---|---|---|---|---|---|
| Burr | 0.000012 | 1.619801 | 1.620137 | 2.065175 | 2.087166 |
| Pareto | 0.000001 | 0.528770 | 0.528773 | 1.444789 | 1.449654 |
| Lognormal | 0.000193 | 18.288933 | 18.473724 | 3.663488 | 3.685818 |
| Weibull | 0.001035 | 99.583679 | 100.61869 | 2.561514 | 2.555888 |

**Table 3.** Comparison of queues using different service time approximations.

| Service time | $\rho$ | $E[S]$ | $E[W]$ | $E[T]$ | $E[N]$ | $E[N_q]$ |
|---|---|---|---|---|---|---|
| Pareto | 0.238095 | 0.476190 | 0.284091 | 0.760281 | 0.380141 | 0.142045 |
| BPH | 0.230504 | 0.461009 | 0.213801 | 0.674809 | 0.337405 | 0.106900 |
| HE | 0.126288 | 0.252576 | 0.211307 | 0.463884 | 0.231942 | 0.105654 |
| BPH_HE | 0.246392 | 0.492785 | 0.268821 | 0.761607 | 0.380803 | 0.134410 |
| BPH_HE adj. | 0.238095 | 0.476190 | 0.248287 | 0.724477 | 0.362239 | 0.124143 |

Validation involves comparing key performance metrics like utilization $\rho$, mean service time $E[S]$, mean waiting time $E[W]$, mean sojourn time $E[T]$, number of jobs in the system $E[N]$, and number of jobs in the queue $E[N_q]$ across the considered models: BPH, HE and BPH_HE. The results are presented in Table 3, where we can observe that the BPH_HE distribution outperforms the BPH and the HE models. The final row (BPH_HE adj.) shows results after adjusting the generator matrix of the PH model to match exactly the mean service time $m_1$ of the original heavy-tailed distribution. The adjusted matrix is computed as:

$$A_{\text{scaled}} = \frac{\hat{m}_1}{m_1} A$$

where $\hat{m}_1$ is the mean associated with the PH distribution with generator $A$. This adjustment makes the approximation of the utilization exact (because it depends only on the mean service time), but worsens some other measures.

Figure 6 illustrates some distribution-based analysis, comparing the waiting time (left) and queue length (right) distributions of the original (M/Pareto/1) queue and the approximate (M/PH/1) queue with BPH_HE service time. These visualizations offer insights into the PH approximation for capturing rare but impactful events, long waiting times, and extended queue lengths.

Overall, these results validate the practical utility of approximating service times using a mixture of BPH and HE models, particularly for capturing performance in typical queuing scenarios.

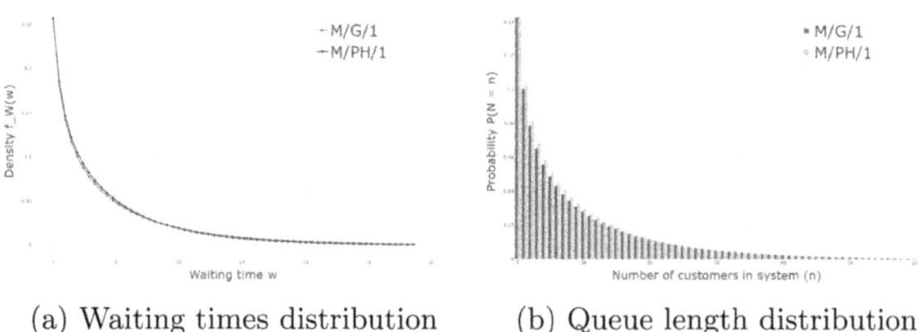


(a) Waiting times distribution          (b) Queue length distribution

**Fig. 6.** Distribution-based comparison.

# 7    Conclusion and Future Work

This research proposed the BPH_HE mixture model to approximate heavy-tailed distributions, addressing the limitations of existing methods by combining the tractability of Bernstein PH distributions with the tail fitting flexibility of hyperexponential models. For the parametrization of the hyperexponential component, a novel initialization method using gradient descent was introduced. The resulting approach accurately captures both body and tail regions of distributions like Burr, Pareto, and Lognormal. Validation through queuing theory confirmed the model's effectiveness in applications. Future work involves applying this heavy-tail aware model in AI algorithms where rare and extreme events play a critical role, such as in generative models or risk-sensitive applications in finance and cybersecurity. With that, we hope to achieve more robust and reliable results compared to traditional Gaussian-based approaches.

# References


1. Foss, S., Korshunov, D., Zachary, S., et al.: An Introduction to Heavy-tailed and Subexponential Distributions, Vol. 6. Springer, Cham (2011). https://doi.org/10.1007/978-1-4614-7101-1
2. Stewart, W.J.: Introduction to the Numerical Solution of Markov Chains. Princeton University Press, Princeton, NJ (1994)
3. Horváth, A., Vicario, E.: Construction of phase type distributions by Bernstein exponentials. In: Iacono, M., Scarpa, M., Barbierato, E., Serrano, S., Cerotti, D., Longo, F. (eds.) Computer Performance Engineering and Stochastic Modelling. EPEW ASMTA 2023 2023. LNCS, vol. 14231, pp. 201–215. Springer, Cham (2023). https://doi.org/10.1007/978-3-031-43185-2_14
4. Horváth, A., Horváth, I., Paolieri, M., Telek, M., Vicario, E.: Approximation of cumulative distribution functions by bernstein phase-type distributions. Perform. Eval. **168** (2025). https://doi.org/10.1016/j.peva.2025.102480
5. Feldman, A., Whitt, W.: Fitting mixtures of exponentials to long-tail distributions to analyze network performance models. Perform. Eval. **31**, 245–279 (1998)
6. Horváth, A., Telek, M.: Approximating heavy tailed behavior with phase-type distributions. In: Proceedings of the 3rd International Conference on Matrix-Analytic Methods in Stochastic models, Leuven, Belgium, pp. 191–214, 2000
7. Neuts, M.: Probability distributions of phase type. In: Liber Amicorum Prof. Emeritus H. Florin, University of Louvain, pp. 173–206, 1975
8. Johnson, M.A., Taaffe, M.R.: Matching moments to phase distributions: mixtures of Erlang distributions of common order. Stoch. Model. **5**(4), 711–743 (1989)
9. Bobbio, A., Horváth, A., Telek, M.: Matching three moments with minimal acyclic phase type distributions. Stoch. Model. **21**(2–3), 303–326 (2005)
10. Horváth, A., Telek, M.: Matching more than three moments with acyclic phase type distributions. Stoch. Model. **23**(2), 167–194 (2007). https://doi.org/10.1080/15326340701300712

11. Buchholz, P., Kriege, J.: A heuristic approach for fitting maps to moments and joint moments. In: Proceedings of the 2009 Sixth International Conference on the Quantitative Evaluation of Systems, QEST '09, IEEE Computer Society, USA, pp. 53–62, 2009. https://doi.org/10.1109/QEST.2009.36

12. Sherzer, E., Resheff, Y., Telek, M.: An unconstrained optimization approach to moment fitting with phase type distributions (2025). arXiv:2505.20379

13. Horváth, A., Telek, M.: Phase Type Distributions: Theory and Application, Wiley-ISTE, Hoboken, 2024

14. Asmussen, S., Nerman, O., Olsson, M.: Fitting phase-type distributions via the EM algorithm. Scand. J. Stat. **23**(4), 419–441 (1996)

15. Okamura, H., Dohi, T., Trivedi, K.S.: A refined EM algorithm for PH distributions. Perform. Eval. **68**(10), 938–954 (2011)

16. Bobbio, A., Telek, M.: A benchmark for PH estimation algorithms: results for Acyclic-PH. Stoch. Model. **10**, 661–677 (1994)

17. Haddad, S., Moreaux, P., Chiola, G.: Efficient handling of phase-type distributions in generalized stochastic petri nets. In: Azéma, P., Balbo, G. (eds.) ICATPN 1997. LNCS, vol. 1248, pp. 175–194. Springer, Heidelberg (1997). https://doi.org/10.1007/3-540-63139-9_36

18. Neuts, M.F.: Matrix-Geometric Solutions in Stochastic Models: An Algorithmic Approach, Johns Hopkins University Press, Baltimore, 1981. foundational work on matrix-geometric methods for solving stochastic models with phase-type distributions

19. Latouche, G., Ramaswami, V.: Introduction to Matrix Analytic Methods in Stochastic Modeling, ASA-SIAM, Philadelphia, 1999. comprehensive treatment of matrix-analytic methods for phase-type models

20. Cumani, A.: Esp - a package for the evaluation of stochastic Petri nets with phase-type distributed transition times. In: International Workshop on Timed Petri Nets, IEEE Computer Society, USA, pp. 144–151, 1985

21. Horváth, A., Telek, M.: PhFit: a general phase-type fitting tool. In: Field, T., Harrison, P.G., Bradley, J., Harder, U. (eds.) Computer Performance Evaluation: Modelling Techniques and Tools. TOOLS 2002. LNCS, vol. 2324, pp. 82–91. Springer, Berlin, Heidelberg (2002). https://doi.org/10.1007/3-540-46029-2_5

22. Hyperstar: A tool for fitting hyper-erlang distributions, https://www.mi.fu-berlin.de/inf/groups/ag-dds/Tools/HyperStar/index.html, FU Berlin, AG DDS Group. Accessed 08 Aug 2025

23. Horváth, G., Telek, M.: Butools 2: a rich toolbox for markovian performance evaluation. In: Proceedings of the 10th EAI International Conference on Performance Evaluation Methodologies and Tools, ACM, 2017. 10.4108/eai.25-10-2016.2266400

24. Afolalu, S., Ikumapayi, O., Abdulkareem, A., Emetere, M., Adejumo, O.: A short review on queuing theory as a deterministic tool in sustainable telecommunication system. Mater. Today: Proc. **44**, 2884–2888 (2021). international Conference on Materials, Processing and Characterization. https://doi.org/10.1016/j.matpr.2021.01.092

25. Harchol-Balter, M.: Open problems in queueing theory inspired by datacenter computing. Queueing Syst. 3–37 (2021). https://doi.org/10.1007/s11134-020-09684-6

26. Zi-Yi, G.: Heavy-tailed distributions and risk management of equity market tail events. J. Risk Control 4(1) (2017)

27. Piantadosi, S.T.: Zipf's word frequency law in natural language: a critical review and future directions. Psychon. Bull. Rev. **21**(5), 1112–1130 (2014). https://doi.org/10.3758/s13423-014-0585-6

28. Su, S., Zhang, H., Wang, Z.: Evaluating large language models on twitter based on hashtag dynamics and scaling properties. In: Zhang, H., Su, J., Shang, J. (eds.) Intelligent Multilingual Information Processing. IMLIP 2024. CCIS, vol. 2395, pp. 236–248. Springer, Singapore (2025). https://doi.org/10.1007/978-981-96-5123-8_16
29. Kingma, D.P., Ba, J.: Adam: a method for stochastic optimization (2017). https://arxiv.org/abs/1412.6980

# Response Time in a Tandem Network of Two Processor Sharing Queues


Julianna Bor[✉] and Peter G. Harrison

Imperial College London, 180 Queen's Gate, London SW7 2AZ, UK
`{jb1316,pgh}@ic.ac.uk`



**Abstract.** We calculate the response time probability density for two processor-sharing (PS) queues in tandem for the first time. We use the generating functions method to obtain a system of functional equations for the Laplace-Stieltjes transform (LST) of the conditional response time distribution, where the conditioning is on the queue lengths at the time of arrival of the tagged task. Similar equations have been derived previously; however, they remained unsolved. Here, we utilise a numerical approximation technique to find the solution for the conditional LST, whereupon the unconditional LST is obtained directly using the product-form solution of the queue length probabilities. We invert the LST numerically to obtain the response time density itself. Furthermore, an efficient, recursive formula is obtained for the calculation of response time moments. The results are compared against regenerative simulation and show remarkable accuracy. Finally, the PS-PS tandem network is compared against the PS-FCFS and FCFS-FCFS systems with the same parameters.

**Keywords:** Queueing theory · Response times · Generating functions


## 1 Introduction

The most basic open queueing network consists of two $M/M/1$ queues arranged in tandem. That is, the inter-arrival and service times are exponential, and an arriving job joins the first of the two queues and proceeds to the second queue after it is served at the first queue. The queueing system is illustrated in Fig. 1. Despite the simplicity of this model, if one or both queues are under processor-sharing (PS) scheduling, obtaining the response time distribution is not straightforward. The functional equations for the Laplace-Stieltjes transform (LST) of the response time distribution for a PS-FCFS tandem, as well as a PS-PS tandem, are obtained in [5]; In both cases, the problem is first reduced to a boundary value problem, which is reduced to a Fredholm equation of the first kind. These are commonly more difficult to solve than those of the second kind. Consequently, the numerical inversion to obtain explicit solutions for the functional equations of the LST of response time distributions were unsuccessful, and the problems remained unsolved. In the case of a PS-FCFS tandem, the first two moments


© The Author(s), under exclusive license to Springer Nature Switzerland AG 2026
L. Carnevali and J. Doncel (Eds.): EPEW 2025, LNCS 15657, pp. 72–85, 2026.
https://doi.org/10.1007/978-3-032-16345-5_6

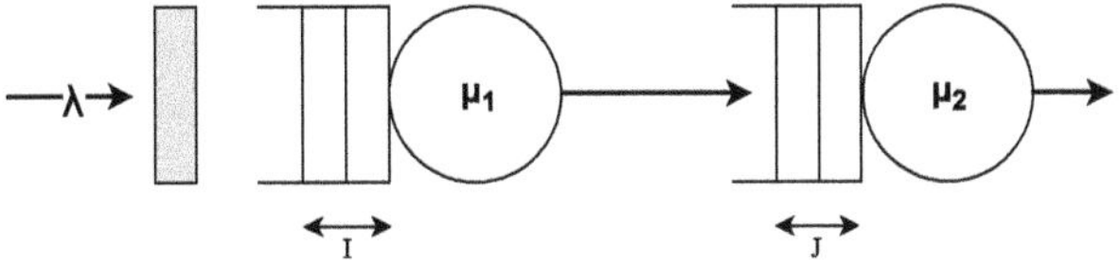


**Fig. 1.** Two PS queues in tandem with Poisson arrivals at rate $\lambda$ and service rates $\mu_1, \mu_2$.

are approximated in [4], and more recently, [7] obtained the LST of the response time distribution for this model using a novel numerical approximation method. The results are compared against simulation and appear to be accurate.

In this paper, we calculate the response time density for a tandem pair of PS queues, for the first time, to our knowledge. We follow similar steps to [5] to derive the functional equations; however, we perform a linear transformation of the generating functions first—to aid numerical accuracy during the solution process—and derive equations in terms of these transformed generating functions. We then solve these equations using the approximation method introduced in [7]. Furthermore, a recursive formula for the efficient calculation of the moments of the response time distribution is derived. The results are compared against regenerative simulation and show very high accuracy. The rest of the paper is organised as follows. In Sect. 2, the functional equations are obtained; these are solved in Sect. 3. Results are evaluated in Sect. 4 and a summary is given in Sect. 5

## 2 Obtaining the Functional Equations

We start by obtaining the functional equations for the generating function of the LST of the conditional response time distribution. Let $\lambda$ denote the arrival rate and $\mu_1, \mu_2$ the service rates at queues 1 and 2, respectively. It is known that if $\lambda < \min(\rho_1, \rho_2)$, a stationary queue-length distribution exists and has a product form, given by the formula

$$\pi_{ij} = (1 - \rho_1)(1 - \rho_2)\rho_1^i \rho_2^j$$

where $\rho_1 = \lambda/\mu_1, \rho_2 = \lambda/\mu_2$.

Let $\Phi_1(i, j, s)$ denote the LST of the remaining conditional response time distribution of a task at queue 1 when the queue lengths are $I = i$ and $J = j$ at queues 1 and 2, respectively, and define the corresponding generating function

$$G(x, y, s) = \sum_{i=0}^{\infty} \sum_{j=0}^{\infty} \Phi_1(i, j, s) x^i y^j$$

Analogously, for a task at queue 2, we denote the conditional LSTs by $\Phi_2(i, j, s)$ and the corresponding generating function by $F(x, y, s)$. Due to the product form stationary queue-length probabilities, the LST of the response time distribution is straightforwardly calculated as $W(s) = (1 - \rho_1)(1 - \rho_2)G(\rho_1, \rho_2, s)$. Functional equations for generating functions $G$ and $F$ are derived in [5], however instead of using these equations directly, we apply a simple transformation to the generating functions and derive similar equations for these new functions. The transformations enhance the accuracy of the numerical approximation when obtaining the solutions of the equations in Sect. 3. The rationale behind this is as follows: when $s = 0$, the LST of the remaining sojourn time of a task at queue 1, $\Phi_1(i, j, s)$, simplifies to the integral of the density function and therefore $\Phi_1(i, j, 0) = 1$ for all $i, j$. Furthermore, for $\Re(s) > 0$, $|\Phi_1(i, j, s)| < 1$, $i, j \geq 0$ as

$$|\Phi_1(i, j, s)| \leq \int_0^\infty f(t)|e^{-st}|dt = \int_0^\infty f(t)e^{-\Re(s)t}dt < 1$$

Of course, the same applies to $\Phi_2(i, j, s)$. Thus, $G(x, y, 0) = F(x, y, 0) = \frac{1}{(1-x)(1-y)}$ and for $\Re(s) > 0$, $|G(x, y, s)| < \frac{1}{(1-|x|)(1-|y|)}$. We define

$$H(x, y, s) = (1 - x)(1 - y)G(x, y, s)$$

and

$$J(x, y, s) = (1 - x)(1 - y)F(x, y, s)$$

Then we have $H(x, y, 0) = J(x, y, 0) = 1$. These functions are numerically more desirable than $G(x, y, s)$ and $F(x, y, s)$ with their sharp peaks near $x = 1$ and $y = 1$. The LST is now given by $W(s) = H(\rho_1, \rho_2, s)$.

The following proposition states the functional equations for $H$ and $J$. The omitted steps can be found in the analogous proofs for $G$ and $F$ in [5].

**Proposition 1.** *The generating functions $H(x, y, s)$ and $J(x, y, s)$ are given by the following equations.*

$$
\begin{aligned}
H(x,y,s)&\left(\mu_2 y + \frac{sy - \mu_1 x + \mu_1 y}{(1 - y)}\right) \\
&- \frac{\partial H(x, y, s)}{\partial x}\left(\frac{(1 - x)\left[x(x - y)\mu_1 + y(\lambda - x(s + \lambda)) - x(1 - y)\mu_2)\right]}{1 - y}\right) \\
&+ H(x, 0, s)(\mu_1 x - \mu_2 y) + \frac{\partial H(x, 0, s)}{\partial x}(1 - x)x(\mu_1 x - \mu_2 y) \\
&= \frac{\mu_1(1 - x)y}{(1 - y)}J(x, y, s)
\end{aligned}
\tag{1}
$$

$$
\begin{aligned}
J(x,y,s)&\left(\frac{\lambda(1 - 1/x) + s + \mu_1(1 - x)}{1 - y} + \mu_2\right) \\
&+ \frac{\partial}{\partial y}J(x, y, s)\left(y[\lambda(1 - 1/x) + \mu_2(1 - y) + s] + \mu_1(y - x)\right) \\
&+ J(0, y, s)\frac{1 - x}{1 - y}\left(\frac{\lambda}{x} - \mu_1\right) + \frac{\partial}{\partial y}J(0, y, s)(1 - x)y\left(\frac{\lambda}{x} - \mu_1\right) = \mu_2
\end{aligned}
\tag{2}
$$

*Proof.* We start with a task at queue 2. A recurrence formula is derived for the density of conditional remaining response time at time $t$ by expressing it as the sum of $h$—an infinitesimal initial interval—and the remaining conditional response time at time $t + h$. The resulting equation in terms of the conditional LSTs is [5][Equation (2.3)]

$$\Phi_2(i, j, s)\left(\lambda + \mu_1 I_{i>0} + \mu_2 + s\right) = \lambda\Phi_2(i + 1, j, s)$$

$$+\mu_1 I_{i>0}\Phi_2(i - 1, j + 1, s) + \mu_2\frac{j}{j + 1}\Phi_2(i, j - 1, s) + \frac{\mu_2}{j + 1}$$

Multiplying by $(j + 1)x^i y^j$ and summing for all $i$ and $j$ values, after repeated but routine transformations of the resulting double sums, we get

$$F(x, y, s)\left(\lambda\left(1 - \frac{1}{x}\right) + \mu_1 + \mu_2(1 - y) + s\right)$$

$$+ \frac{\partial}{\partial y}F(x, y, s)\left(y\left(\lambda\left(1 - \frac{1}{x}\right) + \mu_2(1 - y) + s\right) + \mu_1(y - x)\right)$$

$$+ F(0, y, s)\left(\frac{\lambda}{x} - \mu_1\right) + \frac{\partial}{\partial y}F(0, y, s)\left(y\left(\frac{\lambda}{x} - \mu_1\right)\right) = \frac{\mu_2}{(1 - x)(1 - y)}$$

Multiplying both sides by $(1 - x)(1 - y)$ and using the definition of $J(x, y, s)$ to replace the $F$ terms, after some rearrangement we have

$$J(x, y, s)\left(\frac{\lambda(1 - 1/x) + s + \mu_1(1 - x)}{1 - y} + \mu_2\right)$$

$$+ \frac{\partial}{\partial y}J(x, y, s)\left(y[\lambda(1 - 1/x) + \mu_2(1 - y) + s] + \mu_1(y - x)\right)$$

$$+ J(0, y, s)\frac{1 - x}{1 - y}\left(\frac{\lambda}{x} - \mu_1\right) + \frac{\partial}{\partial y}J(0, y, s)(1 - x)y\left(\frac{\lambda}{x} - \mu_1\right) = \mu_2$$

Moving on to a task at queue 1, by the same reasoning, the recurrence formula, in this case, is [5][Equation (2.5)]

$$\Phi_1(i, j, s)\left(\lambda + \mu_1 + \mu_2 I_{j>0} + s\right) = \lambda\Phi_1(i + 1, j, s)$$

$$+\mu_1\left(\frac{i}{i + 1}\Phi_1(i - 1, j + 1, s) + \frac{1}{i + 1}\Phi_2(i, j, s)\right) + \mu_2\Phi_1(i, j - 1, s)$$

Multiplying by $(i + 1)x^i y^j$ and summing for all $i$ and $j$ values, the resulting equation, after some transformation, is

$$G(x, y, s)\left(\lambda + \mu_1\left(1 - \frac{x}{y}\right) + \mu_2(1 - y) + s\right)$$

$$+ \frac{\partial}{\partial x}G(x, y, s)\left(\lambda(x - 1) + \mu_1 x\left(1 - \frac{x}{y}\right) + \mu_2 x(1 - y) + xs\right)$$

$$+ G(x, 0, s)\left(\mu_1\frac{x}{y} - \mu_2\right) + \frac{\partial}{\partial x}G(x, 0, s)\left(x\left(\mu_1\frac{x}{y} - \mu_2\right)\right) = \mu_1 F(x, y, s)$$

Multiplying by $(1-x)^2 y$ and using the definitions of $H(x,y,s)$ and $J(x,y,s)$ to replace the $G$ and $F$ terms, respectively, after some transformation, we have

$$H(x,y,s)\left(\mu_2 y + \frac{sy - \mu_1 x + \mu_1 y}{(1-y)}\right)$$
$$- \frac{\partial}{\partial x}H(x,y,s)\left(\frac{(1-x)\left[x(x-y)\mu_1 + y(\lambda - x(s+\lambda)) - x(1-y)\mu_2)\right]}{1-y}\right)$$
$$+ H(x,0,s)\left(\mu_1 x - \mu_2 y\right) + \frac{\partial}{\partial x}H(x,0,s)(1-x)x\left(\mu_1 x - \mu_2 y\right)$$
$$= \mu_1 \frac{y(1-x)}{1-y}J(x,y,s)$$

Note, that while (2) has only $J$-terms, (1) has both $H$- and $J$-terms. Therefore, a solution must first be obtained for (2) which is then plugged into (1), after which it can be solved.

## 3  Solving the Functional Equations

The system of functional equations obtained in Proposition 1 is solved by a numerical approximation technique that transforms the equations into a system of linear equations. The method is first introduced in [7] and is explained in detail via a simple example in [3]. Here we reproduce only Fig. 5 and Table 6, and provide a brief summary of the method for the reader's convenience.

Instead of trying to solve $H(x,y,s)$ and $J(x,y,s)$ for all $x,y$ in the complex plane—treating the Laplace variable $s$ as a parameter—the idea is to first restrict ourselves to a fixed number of input point pairs placed equidistantly around circles in the complex plane in two dimensions. Suppose that we know the value of the generating functions at these $(x,y)$ point pairs. Then, we could use interpolation to estimate the functions everywhere on the boundaries of the circles, and apply Cauchy's integral theorem [8,9] to estimate them inside the circles. Figure 5 visualises these steps in one dimension when the number of points around the circle is set to 8.

We can think of the values of the generating functions at the point pairs as a single vector of unknowns, where each element represents the value of the generating function at a specific point pair. By examining the operations in the functional equations, the goal is to find a matrix representation for each operation that acts on this vector. Table 6 lists all the required operations along with their corresponding constant matrices; details on how to compute these matrices can be found in [7]. For example, if the operation is multiplication by a known value $c$, the corresponding matrix is diagonal with all diagonal entries equal to $c$. Using the notation of Table 6, if the vector of unknowns is $\boldsymbol{f} = (f(z_1), \ldots, f(z_n))^\top$, we have $\boldsymbol{\Lambda}(c)\boldsymbol{f} = (cf(z_1), \ldots, cf(z_n))^\top$.

## 3.1   Calculating the LST

The functional Eq. (2) of $J(x, y, s)$ at the chosen discretisation point-pairs, represented by the lexicographically ordered vector $\boldsymbol{j} = (J(x_i, y_j)|1 \leq i \leq m, 1 \leq j \leq n)$, is transformed into the matrix-vector equation

$$\left(\boldsymbol{\Lambda}(\boldsymbol{a}_1) + \boldsymbol{\Lambda}(\boldsymbol{a}_2)\boldsymbol{M}_{0,1} + \boldsymbol{\Lambda}(\boldsymbol{a}_3)\boldsymbol{U}_0 + \boldsymbol{\Lambda}(\boldsymbol{a}_4)\boldsymbol{M}_{0,1}\boldsymbol{U}_0\right)\boldsymbol{j} = (\mu_2, \ldots, \mu_2)^T$$

where $\boldsymbol{a}_1, \boldsymbol{a}_2, \boldsymbol{a}_3, \boldsymbol{a}_4$ are the vector representations of the functions $A_1, A_2, A_3, A_4$ defined by

$$A_1(x, y) = \left(\frac{\lambda(1 - 1/x) + s + \mu_1(1 - x)}{1 - y} + \mu_2\right)$$

$$A_2(x, y) = \left(y[\lambda(1 - \frac{1}{x}) + \mu_2(1 - y) + s] + \mu_1(y - x)\right)$$

$$A_3(x, y) = \frac{1 - x}{1 - y}\left(\frac{\lambda}{x} - \mu_1\right) \qquad A_4(x, y) = (1 - x)y\left(\frac{\lambda}{x} - \mu_1\right)$$

This transformation is performed by matching each operation in (3) to the corresponding matrix representation based on Table 6. For example, the second term of (2) involves differentiation with respect to $y$. Looking up this operation in Table 6, we find that the corresponding matrix is $\boldsymbol{M}_{0,1}$. Define the matrix $\boldsymbol{\Gamma}_J(s)$ as

$$\boldsymbol{\Gamma}_J(s) = \left(\boldsymbol{\Lambda}(\boldsymbol{a}_1) + \boldsymbol{\Lambda}(\boldsymbol{a}_2)\boldsymbol{M}_{0,1} + \boldsymbol{\Lambda}(\boldsymbol{a}_3)\boldsymbol{U}_0 + \boldsymbol{\Lambda}(\boldsymbol{a}_4)\boldsymbol{M}_{0,1}\boldsymbol{U}_0\right)^{-1} \tag{3}$$

Note that as the notation suggests, $\boldsymbol{\Gamma}_J(s)$ depends on the value of the Laplace parameter $s$. The solution for $\boldsymbol{j}$ for a given $s$ value can be calculated as

$$\boldsymbol{j} = \boldsymbol{\Gamma}_J(s)(\mu_2, \ldots, \mu_2)^\top \tag{4}$$

Analogously, (1) is transformed into the matrix-vector equation

$$\left(\boldsymbol{\Lambda}(\boldsymbol{b}_1) - \boldsymbol{\Lambda}(\boldsymbol{b}_2)\boldsymbol{M}_{1,0} + \boldsymbol{\Lambda}(\boldsymbol{b}_3)\boldsymbol{V}_0 + \boldsymbol{\Lambda}(\boldsymbol{b}_4)\boldsymbol{M}_{1,0}\boldsymbol{V}_0\right)\boldsymbol{h} = \boldsymbol{\Lambda}(\boldsymbol{b}_5)\boldsymbol{j}$$

where $\boldsymbol{b}_i, 1 \leq i \leq 5$ are the vector representations of the functions $B_i$ defined by

$$B_1(x, y) = \mu_2 y + \frac{sy - \mu_1 x + \mu_1 y}{(1 - y)}$$

$$B_2(x, y) = \frac{(1 - x)\left[x(x - y)\mu_1 + y(\lambda - x(s + \lambda) - x(1 - y)\mu_2)\right]}{1 - y}$$

$$B_3(x, y) = \mu_1 x - \mu_2 y \qquad B_4(x, y) = (1 - x)x(\mu_1 x - \mu_2 y) \qquad B_5(x, y) = \frac{\mu_1(1 - x)y}{(1 - y)}$$

Define $\boldsymbol{\Gamma}_H(s)$ as

$$\boldsymbol{\Gamma}_H(s) = \left(\boldsymbol{\Lambda}(\boldsymbol{b}_1) - \boldsymbol{\Lambda}(\boldsymbol{b}_2)\boldsymbol{M}_{1,0} + \boldsymbol{\Lambda}(\boldsymbol{b}_3)\boldsymbol{V}_0 + \boldsymbol{\Lambda}(\boldsymbol{b}_4)\boldsymbol{M}_{1,0}\boldsymbol{V}_0\right)^{-1} \tag{5}$$

Again, $\boldsymbol{\Gamma}_H(s)$ depends on the value of the parameter $s$. The solution for $\boldsymbol{h}$ for a given $s$ value can be calculated as

$$\boldsymbol{h} = \boldsymbol{\Gamma}_H(s)\boldsymbol{\Lambda}(b_5)\boldsymbol{j}$$

where $\boldsymbol{j}$ is given by (4).

## 3.2  Calculating the Moments

The moments of the response time can be calculated from the LST by differentiating it with respect to $s$ and evaluating at $s = 0$. Let $m_n$ denote the $n$th moment, then we have

$$m_n = (-1)^n \frac{\partial^n}{\partial s^n} W(0) = (-1)^n \frac{\partial^n}{\partial s^n} H(\rho_1, \rho_2, 0)$$

The following lemmas derive a recursive formula for obtaining the derivatives of $J(x, y, s)$ and $H(x, y, s)$ with respect to $s$ using their vector representation $\boldsymbol{j}$ and $\boldsymbol{h}$, respectively.

**Lemma 1.** *The $k$th partial derivative of $J(x, y, s)$ with respect to $s$ at $s = 0$ has vector representation $\boldsymbol{j}_k$ given by the recurrence*

$$\boldsymbol{j}_0 = (1, \ldots, 1)^{\top}$$
$$\boldsymbol{j}_k = \boldsymbol{\Gamma}_J(0) \left( -k\boldsymbol{\Lambda}\left(\boldsymbol{c}_1\right)\boldsymbol{j}_{k-1} - k\boldsymbol{\Lambda}(\boldsymbol{c}_2)\boldsymbol{j}'_{k-1}\right)$$
$$\boldsymbol{j}'_k = \boldsymbol{M}_{0,1}\boldsymbol{j}_k$$

*where $\boldsymbol{\Gamma}_J(0)$ is given by (3) with substitution $s = 0$ and $\boldsymbol{c}_1, \boldsymbol{c}_2$ are the vector representations of functions $C_1, C_2$ defined as*

$$C_1(x, y) = \frac{1}{1 - y} \qquad C_2(x, y) = y$$

*Proof.* First, note that $J(x, y, 0) = 1$ by construction. Differentiating (2) $k$ times with respect to $s$, we get

$$J^{(k)}(x, y, s) \left( \frac{\lambda(1 - 1/x) + s + \mu_1(1 - x)}{1 - y} + \mu_2 \right)$$
$$+ \frac{\partial}{\partial y} J^{(k)}(x, y, s) \left( y[\lambda(1 - 1/x) + \mu_2(1 - y) + s] + \mu_1(y - x) \right)$$
$$+ J^{(k)}(0, y, s)\frac{1 - x}{1 - y} \left( \frac{\lambda}{x} - \mu_1 \right) + \frac{\partial}{\partial y} J^{(k)}(0, y, s)(1 - x)y \left( \frac{\lambda}{x} - \mu_1 \right)$$
$$= -k \left( \frac{1}{1 - y} J^{k-1}(x, y, s) + y\frac{\partial}{\partial y} J^{(k-1)}(x, y, s) \right)$$

The coefficients on the left-hand side of this equation are the same as in (2), only the vector on the right-hand side changes. Hence, we can reuse $\boldsymbol{\Gamma}_J(s)$ with substitution $s = 0$ from (3) and calculate the $\boldsymbol{j}_k$ vectors very efficiently by updating only the right-hand side.

Once the $\boldsymbol{j}_k$ vectors are calculated, the derivative of $H(x, y, s)$ with respect to $s$ can be calculated using the following lemma.

**Lemma 2.** *The $k$th partial derivative of $H(x, y, s)$ with respect to $s$ at $s = 0$ has vector representation $\boldsymbol{h}_k$ given by the recurrence*

$$\boldsymbol{h}_0 = (1, \ldots, 1)^\top$$
$$\boldsymbol{h}_k = \boldsymbol{\Gamma}_H(s) \left( -k\boldsymbol{\Lambda}(\boldsymbol{d}_1)\boldsymbol{h}_{k-1} - k\boldsymbol{\Lambda}(\boldsymbol{d}_2)\boldsymbol{h}'_{k-1} + \boldsymbol{\Lambda}(\boldsymbol{d}_3)\boldsymbol{j}_k \right)$$
$$\boldsymbol{h}'_k = \boldsymbol{M}_{1,0}\boldsymbol{h}_k$$

*where $\boldsymbol{\Gamma}_H(s)$ is given by (5) with substitution $s = 0$, $\boldsymbol{j}_k$ is given by Lemma 1 and $\boldsymbol{d}_1, \boldsymbol{d}_2, \boldsymbol{d}_3$ are vector representations of functions $D_1, D_2, D_3$ defined as*

$$D_1(x, y) = \frac{y}{1 - y} \qquad D_2(x, y) = \frac{(1 - x)xy}{1 - y} \qquad D_3(x, y) = \mu_1 \frac{y(1 - x)}{1 - y}$$

*Proof.* Again, we start by noting that $H(x, y, 0) = 1$ by construction. Calculating the $k$th derivative of (1) with respect to $s$, we have

$$H^{(k)}(x, y) \left( \mu_2 y + \frac{sy - \mu_1 x + \mu_1 y}{(1 - y)} \right)$$
$$- \frac{\partial H^{(k)}(x, y)}{\partial x} \left( \frac{(1 - x)\left[x(x - y)\mu_1 + y(\lambda - x(s + \lambda)) - x(1 - y)\mu_2)\right]}{1 - y} \right)$$
$$+ H^{(k)}(x, 0) (\mu_1 x - \mu_2 y) + \frac{\partial H^{(k)}(x, 0)}{\partial x}(1 - x)x (\mu_1 x - \mu_2 y)$$
$$+ kH^{(k-1)}(x, y) \left( \frac{y}{1 - y} \right) + k\frac{\partial H^{(k-1)}(x, y)}{\partial x} \left( \frac{(1 - x)xy}{1 - y} \right)$$
$$= \frac{\mu_1(1 - x)y}{(1 - y)} J^{(k)}(x, y, s)$$

Once more, the coefficients on the left-hand side are the same as in (1), only the vector on the right-hand side changes. Hence, we can reuse $\boldsymbol{\Gamma}_H(s)$ with substitution $s = 0$ from (5) and calculate the $\boldsymbol{h}_k$ vectors very efficiently by updating only the right-hand side.

## 4   Numerical Results

As the generating function $H(x, y, s)$ needs to be evaluated at $x = \rho_1, y = \rho_2$ to express the LST of the response time distribution, we choose the vector representations $\boldsymbol{h}$ and $\boldsymbol{j}$ to correspond to point pairs on circles centred at the origin with radii $r_1 = \rho_1, r_2 = \rho_2$ in the complex plane. Therefore, $H(\rho_1, \rho_1, s)$ is calculated from $\boldsymbol{h}$ by selecting the index corresponding to the discretisation point pair $(\rho_1, \rho_2)$, or if the point falls within two discretisation points, by linear interpolation. We use $n = 64$ points in both dimensions and compare the results to simulation. The simulation runs for $5 \times 10^5$ regeneration cycles for three sets

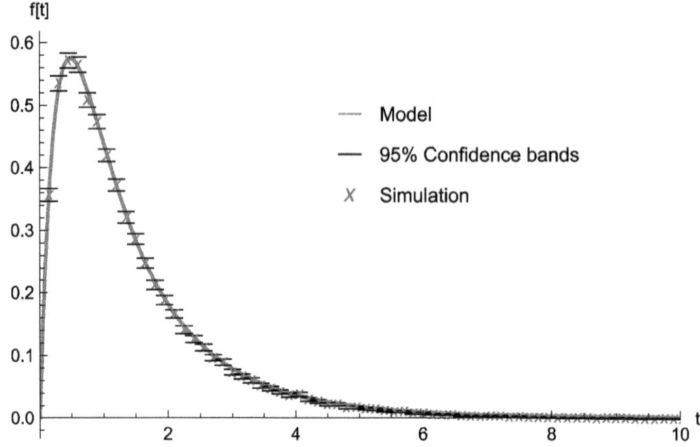


**Fig. 2.** Response time density for a two-server PS tandem network with parameters $\lambda = 1$, $\mu_1 = 2$, $\mu_2 = 3$.

**Table 1.** Response time moments for two PS queues in tandem with parameters $\lambda = 1$, $\mu_1 = 2.$, $\mu_2 = 3$.

| Moment | PS-PS model | Simulation | 95% CB |
|---|---|---|---|
| 1st | 1.50000 | 1.50137 | $\pm$ 0.00542 |
| 2nd | 4.36580 | 4.38277 | $\pm$ 0.04519 |
| 3rd | 21.58840 | 21.78040 | $\pm$ 0.50304 |
| 4th | 163.547 | 165.567 | $\pm$ 7.65520 |

of parameters. The first two sets are chosen in accordance with [7], allowing us to compare our results with a PS–FCFS tandem using the same parameters. These are: a light load case with parameters $\lambda = 1, \mu_1 = 2, \mu_2 = 3$ resulting in a 50% utilisation at the first queue and $\approx 34\%$ at the second queue, and a heavy load case with parameters $\lambda = 1, \mu_1 = 1.4, \mu_2 = 1.5$ giving utilisations $\approx 71\%$ and $\approx 67\%$ at queues 1 and 2, respectively. In addition, we also run a very heavy load case with parameters $\lambda = 2, \mu_1 = 2.4, \mu_2 = 2.6$ resulting in utilisations 83.34% and 77% at the two queues. This stresses the method, since higher utilization leads to less smooth generating functions that are harder to approximate. In all cases, we use the Fixed Talbot method to invert the LST and obtain the probability density function [1].

Figures 2,3 and 4 show the obtained density together with the simulation-generated relative frequency histogram with 95% confidence bands. The results are remarkably accurate with most of the model points falling within the confidence intervals, even in the very heavy load case with utilisations 83.34% and 77% at the two queues.

Next, we calculate the moments using the recursive method given by Lemma 2. Note that this is much faster than calculating the density functions, since for

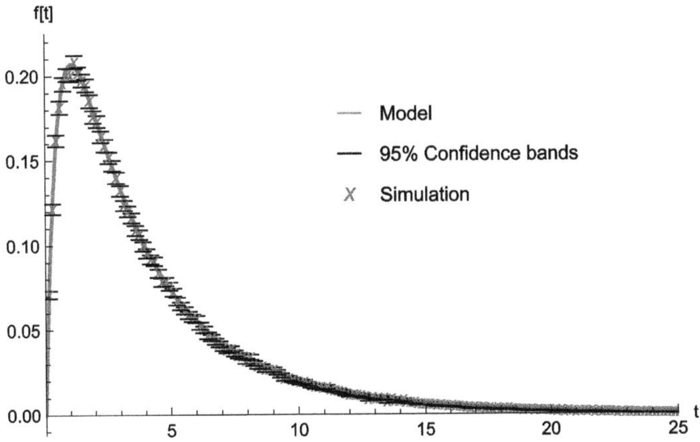

**Fig. 3.** Response time density for a two-server PS tandem network with parameters $\lambda = 1$, $\mu_1 = 1.4$, $\mu_2 = 1.5$.

**Table 2.** Response time moments for two PS queues in tandem with parameters $\lambda = 1$, $\mu_1 = 1.4$, $\mu_2 = 1.5$

| Moment | PS-PS model | Simulation | 95% CB |
|---|---|---|---|
| 1st | 4.50000 | 4.49004 | ± 0.01689 |
| 2nd | 42.98960 | 42.77410 | ± 0.43598 |
| 3rd | 725.31300 | 720.12600 | ± 15.18100 |
| 4th | 18991.800 | 18821.00000 | ± 728.230 |

the moments we only need to calculate $\Gamma_H(s)$ and $\Gamma_J(s)$ at $s = 0$ to obtain all four moments for a given set of parameters, whereas the density function requires the LST to be evaluated at 32 different $s$-values to perform the inversion at a single point. Tables 1,2 and 3 display the first four moments calculated by our model, together with the moments calculated from simulation with the corresponding 95% confidence bands. In the cases of heavy load and very heavy load, we increase the number of discretisation points to $n = 120$ to calculate the moments. All the moments calculated by the model fall within the confidence intervals, except the fourth moment in Table 3, which is slightly outside the 95% confidence interval. However, we know that the first moment must be 1.5, 4.5 and 4.16667 for the three cases, respectively, since these are equal to the first moment in an FCFS-FCFS tandem network. The first moment calculated by the model matches these numbers exactly, whereas the simulation is slightly off. Therefore, we conclude that the higher moments calculated by the model are likely to be more accurate than those of the simulation.

Finally, we compare the first four moments of the PS-PS tandem model to a PS-FCFS and an FCFS-FCFS tandem model with the same parameters. The numbers for the PS-FCFS and FCFS-FCFS models are taken from [7]. The

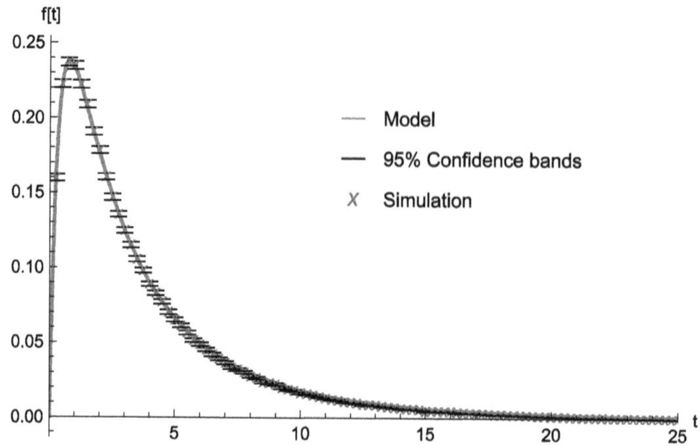



**Fig. 4.** Response time density for a two-server PS tandem network with parameters $\lambda = 2$, $\mu_1 = 2.4$, $\mu_2 = 2.6$.

**Table 3.** Response time moments for two PS queues in tandem with parameters $\lambda = 2$, $\mu_1 = 2.4$, $\mu_2 = 2.6$

| Moment | PS-PS model | Simulation | 95% CB |
|---|---|---|---|
| 1st | 4.16667 | 4.16240 | $\pm$ 0.01833 |
| 2nd | 39.75420 | 39.85860 | $\pm$ 0.47209 |
| 3rd | 697.2140 | 711.71200 | $\pm$ 17.62730 |
| 4th | 19311.60 | 20559.80000 | $\pm$ 992.76300 |

results are given in Tables 4 and 5. As expected, apart from the first moment, all higher moments increase when going from an FCFS to a PS schedule, with PS-PS producing the highest numbers, so that users experience greater variability. Note, however, that the increased variability of PS scheduling is compensated by the fairness of the schedule as the expected time a task spends in the system is proportional to its own service requirement [2]. As a result, smaller tasks do not get stuck behind large tasks as in a FCFS queue. This inherent fairness is why PS tends to be favoured over FCFS.

## 5   Summary

We have obtained functional equations for the transformed generating functions of the LST of conditional response time distribution in a PS-PS tandem network and then applied a numerical approximation technique to solve the resulting system of functional equations. Additionally, we derived a recursive formula

**Table 4.** Comparing response time moments in PS-PS, PS-FCFS and FCFS-FCFS tandems with parameters $\lambda = 1$, $\mu_1 = 2.$, $\mu_2 = 3$.

| Moment | PS-PS model | PS-FCFS | FCFS-FCFS |
|---|---|---|---|
| 1st | 1.50000 | 1.49999 | 1.50 |
| 2nd | 4.36580 | 4.25421 | 3.50 |
| 3rd | 21.58840 | 20.5703 | 11.25 |
| 4th | 163.547 | 149.59000 | 46.50 |

**Table 5.** Comparing response time moments in PS-PS, PS-FCFS and FCFS-FCFS tandems with parameters $\lambda = 1$, $\mu_1 = 1.4$, $\mu_2 = 1.5$

| Moment | PS-PS model | PS-FCFS | FCFS-FCFS |
|---|---|---|---|
| 1st | 4.50000 | 4.49954 | 4.50 |
| 2nd | 42.75710 | 38.53280 | 30.50 |
| 3rd | 708.37300 | 574.1810 | 276.75 |
| 4th | 17863.300 | 13833.000 | 3151.50 |

for computing the response time moments efficiently using the numerical approximation method. Numerical results demonstrate that the density function and the first four moments calculated by the model exhibit very high accuracy when compared to regenerative simulation.

Future work will consider more than two PS queues in tandem. Obtaining exact functional equations for more than two queues quickly becomes impractical as the number of queues increases. However, it might be possible to derive approximate solutions by considering two queues at the time, assuming Poisson arrivals elsewhere and applying a reversed process argument similar to [6].

**Disclosure of Interests.** The authors have no competing interests to declare that are relevant to the content of this article.

## A     Figures and Tables for the Numerical Approximation Algorithm

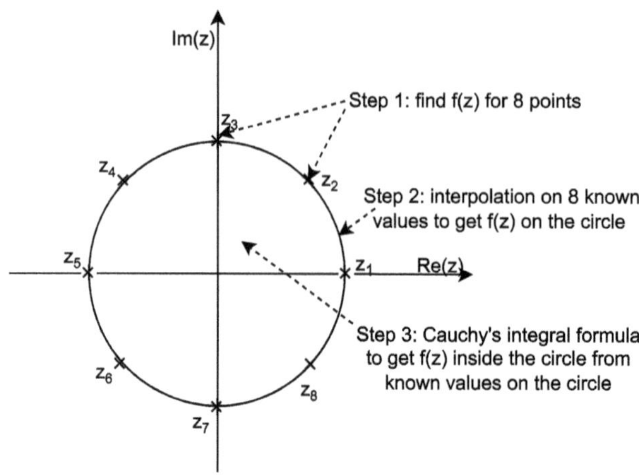


**Fig. 5.** Steps to approximate f(z) from 8 known values around the unit circle.

**Table 6.** Matrices representing operations in the summands of the functional equation for $f(x, y)$.

| Operation | Matrix |
| --- | --- |
| Identity operator giving $f$ | $\boldsymbol{I}^{nm \times nm}$ |
| Multiplication by known function $A(x, y)$ | $\boldsymbol{\Lambda}(A)$ |
| Differentiation wrt $x$ | $\boldsymbol{M}_{1,0}$ |
| Differentiation wrt $y$ | $\boldsymbol{M}_{0,1}$ |
| Evaluation at an internal point $u_0 \in D_x$ | $\boldsymbol{U}_{u_0}$ |
| Evaluation at an internal point $v_0 \in D_y$ | $\boldsymbol{V}_{v_0}$ |
| Partialization giving $f^{\leq}$ | $\boldsymbol{P}^{\leq}$ |
| Partialization giving $f^{\geq}$ | $\boldsymbol{P}^{\geq}$ |

## References


1. Abate, J., Whitt, W.: Numerical inversion of Laplace transforms of probability distributions. ORSA J. Comp. **7**(1) (1995)
2. Asare, B.K., Foster, F.G.: Conditional response times in the M/G/1 processor-sharing system. J. Appl. Probab. **20**(4) (1983)
3. Bor, J., Harrison, P.G.: Response time in a pair of processor sharing queues with Join-the-Shortest-Queue scheduling. In: MASCOTS 2024, pp. 1–8 (2024)

4. Charles, K., Charles, T.: Approximations to the moments of the sojourn time in a tandem queue with overtaking. Stoch. Model. **6**(3), 499–524 (1990)
5. Coffman, E., Jr., Fayolle, G., Mitrani, I.: Sojourn times in a tandem queue with overtaking: reduction to a boundary value problem. Stoch. Model. **2**(1), 43–65 (1986)
6. Harrison, P.G., Bor, J.: Response time distribution in a tandem pair of queues with batch processing. J. ACM **68**(4) (2021)
7. Harrison, P.G.: On the numerical solution of functional equations with application to response time distributions. Appl. Math. Comput. **472** (2024)
8. Hormander, L.: An Introduction to Complex Analysis in Several Variables. Elsevier, Amsterdam (1973)
9. Priestley, H.A.: Introduction to Complex Analysis, second edition (2nd ed.). Oxford University Press, Oxford (2003)

# Performance Analysis and Load Balancing in a Multi-tier Buffered Cellular Network

Taisiia Morozova$^{(\boxtimes)}$ and Ingemar Kaj

Uppsala University, Box 480, 751 06 Uppsala, Sweden
`taisiia.morozova@math.uu.se`

**Abstract.** Multi-tier cellular networks able to handle multiple classes of base stations with varying transmission power and coverage area offer a promising approach to meet increasing traffic demands by proper balancing of the input load. This paper investigates the performance of buffered multi-tier networks with randomly placed transmission nodes. Under simplifying model assumptions of Rayleigh fading and unbounded attenuation, we propose an approach to derive the multi-tier coverage probability and Shannon capacity of typical cells. The goal is to study the trade-off between base station density and interference effects in multi-tier compared to single-tier networks by assessing the system performance under varying policies. We also use stochastic simulation to verify the theoretical results and visualize the system behavior under different parameter settings.

**Keywords:** cellular network · multi-tier · stochastic geometry · performance evaluation · load-balancing · stochastic simulation

## 1 Introduction

Modern communication systems experience growing traffic demands and remain vulnerable to congestion when data rates exceed network capacity. This can lead to significant data loss and transmission delays, making efficient management more challenging. Load balancing is a mechanism to improve network behavior under heavy traffic loads that relies on a suitable strategy to efficiently divide incoming work between serving nodes. Modern ultra-dense 5G networks improve performance via *microcell densification*, employing heterogeneous architectures with macro, micro, and potentially pico cells, differentiated by power and path loss characteristics. Unlike single-tier networks, multi-tier systems benefit from *open access cell association*, where the user equipment (UE) connects to the closest base station (BS). With this policy, a larger number of users will be assigned to densely deployed microcells, facilitating traffic offloading from regular macrocells. Multi-tier deployment of additional, low-cost BSs raises reliability concerns due to lower transmission power and potential data loss in microcells. However, the resulting increase in co-channel downlink interference remains modest as traffic accumulates across tiers. The objective of the multi-tier policy is to improve


L. Carnevali and J. Doncel (Eds.): EPEW 2025, LNCS 15657, pp. 86–100, 2026.
https://doi.org/10.1007/978-3-032-16345-5_7

coverage and data rates by balancing the load among tiers and optimizing the throughput of low-power BSs. In this work, we study a spatial stochastic model of a heterogeneous buffered multi-tier cellular network, suitably simplified to allow for explicit performance evaluation. The aim is to capture some important features of a physical configuration of reliable network nodes, comparing pure-loss, no-loss, and limited-loss settings, as well as evaluating the performance of single-tier and multi-tier networks. Specifically, we analyze the performance of the multi-tier networks in terms of coverage probability and Shannon capacity and briefly discuss their advantages over single-tier networks. Finally, we use stochastic simulation to validate the theoretical findings.

## 1.1 Previous and Related Work

Systems with multi-tier deployment are considered, among other works, in [2–4, 8,9], where the main focus is on the networks with macro and micro BSs, with the overall goal of achieving better coverage through load balancing. In particular, in [3], the authors examine a downlink system featuring multiple classes, where the nodes remain continuously active, and provide analytical formulas for both coverage probability and channel capacity. A similar system is considered in [4], where the authors mainly study load balancing and performance evaluation.

In this study, we analyze a stochastic geometric model of a multi-tier cellular network with a retransmission policy, focusing on coverage probability and Shannon capacity. The approach extends our previous work [6], which introduces a single-tier model. In this previous study, BSs deployed via the Poisson-Voronoi tessellation have transmitters equipped with buffers for failed transmissions. A crucial element in this approach is the activity of the nodes that arises from the inherent structure of the buffer. In [6], we find the activity probability of a node required for workload management and system stability by solving a non-linear balance equation. In [7], we further examine the channel capacity of the system in terms of the Shannon formula.

## 1.2 Our Contribution

In this work, we further develop methodologies presented in [6] together with concepts from queuing theory, and derive analytical expressions for coverage probability and capacity of each network tier. We propose a load-balancing mechanism across multiple tiers that is able to capture some of the complex features of the multi-tier system. Our goal is to evaluate relevant performance metrics and demonstrate that the multi-tier approach results in higher coverage and capacity compared to a single-tier. The paper is organized as follows. In Sect. 2, we define a model of a multi-tier buffered cellular network and derive the coverage probability for a multi-tier network as a function of activity probability. Section 3 presents the main results, including the derivation of the activity probability and Shannon capacity. In Sect. 4, we discuss the multi-tier model where each tier has a different path loss exponent, and give some numerical results. Finally, in Sect. 5, a simulation study and validation of some analytical expressions are provided.

## 2  Multi-tier System Model

We consider a heterogeneous network with one class of users and $J \geq 1$ classes (tiers) of BS, located in the $d$-dimensional space $R^d$. From a physical perspective, these classes correspond to distinct types of antenna (e.g. microcells, femtocells) characterized by differences in coverage radius and signal power. Normally, macrocells are equipped with BSs with signal power higher than that of microcells or picocells. Data transmission follows a downlink traffic pattern, where BSs deliver data to UEs within their designated transmission cells. We assume that the positions of the sending and receiving nodes are generated by independent spatial Poisson point processes (PPP) in $\mathbb{R}^d$ of different intensities. Let $\Phi = \{\Phi^1, \ldots, \Phi^J\}$, $\Phi^i \sim \mathrm{PPP}(\lambda_i)$, be the collection of independent PPPs representing the locations of BS's of all tiers. Also, let $\Phi_{\mathrm{UE}} \sim \mathrm{PPP}(\lambda_0)$ be another Poisson point process with intensity $\lambda_0$ that represents the spatial locations of the users. For any $j = 1, \ldots, J$, we assume that $\Phi^j, \Phi_{\mathrm{UE}}$ are independent. The total BS intensity is $\lambda = \sum_{j=1}^{J} \lambda_j$, and the bandwidth of the network is $w = \lambda_0/\lambda$, which is the average number of users per BS. In each time interval, any BS of the $i$-th tier obtains a new signal with probability $p$, which is hence the input rate of the network, and transmits it with signal power $S^{(i)} \sim \exp(\mu_i)$, $1 \leq i \leq J$. It is worth noting that the exponential transmit power model is widely used in the analysis of cellular networks, as it captures realistic line-of-sight (LoS) path-loss characteristics together with Rayleigh fading effects [5]. To connect the users with their target BSs, we employ the association principle that each user connects to the BS of any tier at the shortest Euclidean distance, regardless of transmit power. As a result, all users are divided into cells of a Poisson-Voronoi tessellation with intensity $\lambda$, such that an average cell has volume $1/\lambda$. The association probability that a typical receiver is associated with a BS of tier $i$ is

$$\alpha_i = \lambda_i \Big/ \sum_{j=1}^{J} \lambda_j, \quad i = 1, \ldots, J. \tag{1}$$

We consider a typical UE node in an arbitrary time slot and assume, without loss of generality, that this user is located at the origin. By construction of the network cells, the UE at the origin belongs to a cell served by a tier-$i$ BS with probability $\alpha_i$. Within this target cell, it follows from the properties of spatial Poisson processes that the random position $X$ of the BS has the distribution

$$\mathbb{P}(X \in B) = \lambda \int_B e^{-\lambda |B(0,1)| \|x\|^d} \, dx, \quad B \subset \mathbb{R}^d. \tag{2}$$

Let $\ell_0$ be an auxiliary parameter that describes the radius of a ball in $\mathbb{R}^d$ with the same volume as a cell, $|B(0,1)|\ell_0^d = 1/\lambda$, where $|B(0,1)|$ is the volume of the unit ball. Attenuation of a signal occurs with (tier-independent) path loss exponent $\beta > d$, such that the power received at $x \in \mathbb{R}^d$ of a unit signal transmitted at the origin is

$$a_\beta(x) = |x/\ell_0|^{-\beta} \tag{3}$$

is a properly scaled attenuation function. As downlink traffic arrives with probability $p$ per slot in all BSs, receiving nodes experience interference from simultaneous transmissions across all levels, leading to potential failures. To mitigate the loss, each BS has a buffer in which failed signals are stored for retransmission in the next slot, with no limit on the number of attempts. A signal remains in the buffer until it is delivered successfully. The presence of the buffer, along with the input rate $p$, makes it necessary to distinguish between the nodes that currently handle the signals and those that do not have a job to do. To address this, we use the notation of *active nodes* (or *busy nodes*) as those that handle a new or buffered signal. We assume that each BS of any tier has a buffer of capacity $K$ beyond which signals are lost. In most physical systems, it is natural to have a buffer of finite capacity $K < \infty$. This setting means that if, at a given time slot, a BS of any tier has failed to deliver the signal and its respective buffer is full, the signal is permanently lost. However, certain network models, for example, the well-known Jackson networks, assume infinite capacity of buffers. In wireless communications, infinite buffers are useful for studying system delays and performance without concern about packet loss, and this model provides guidance on how well scheduling and access methods work when buffers are large.

Hence, in our analysis, we account for the case $K = \infty$, which is also addressed as a case of a no-loss system. Let $B^{(i)} \leq K$ be the number of buffered signals in a typical BS in tier $i$. The *activity probability* of a BS in tier $i$ is given by

$$q_i = 1 - (1-p)\mathbb{P}(B^{(i)} = 0), \quad i = 1,\ldots,J. \tag{4}$$

In other words, a typical transmitter is idle whenever it has no new signal and has an empty buffer, and is active (busy) otherwise. In the simplest case of a *pure loss* system, when $K = 0$ and the base stations lack the capacity to store signals in a buffer, then $B^{(i)} = 0$ and therefore $q_i = p$ for all $i$.

For now we fix $\bar{q} = (q_1,\ldots,q_J)$. Equation (4) connects a conditional buffer size distribution with the given input rate $p$ and the busy server parameters in $\bar{q}$. To determine whether a signal from a fixed $i$-th tier BS is successfully transmitted, we use a classical signal-to-interference-and-noise ratio (SINR) given by

$$\mathrm{SINR}^{(i)}(x) = \frac{a_\beta(x)S_x^{(i)}}{\mathcal{I}^{(i)}(x) + \sigma^2},$$

where $x$ is the cell location of the BS, $S_x^{(i)}$ is the corresponding signal power, exponentially distributed with mean $1/\mu_i$, and $\sigma^2$ is a constant noise contribution. The network interference experienced by the transmitter at distance $x$ from the origin is defined as

$$\mathcal{I}^{(i)}(x) = \sum_{j=1}^{J} \sum_{y \in \Phi^j \setminus x} a_\beta(y) S_y^{(j)} \mathbf{1}_{\{\text{transmitter from tier } j \text{ is active}\}}$$

$$= \sum_{j=1}^{J} \sum_{y \in \Phi^j \setminus x} a_\beta(y) S_y^{(j)} Q_y^{(j)}, \tag{5}$$

by adding over all active transmitters in other cells. Here, $\{Q_y^{(j)}\}$ is a family of independent random variables such that $Q_y^{(j)} \sim \text{Bin}(1, q_j)$ and $q_j$ from (4). For simplicity, we consider the case $\sigma^2 = 0$ and write SIR, rather than SINR.

## 2.1  Comparison of Multi-Tier with Single-Tier

The reference case is a single-tier system with only regular BSs, i.e., $\alpha_1 = 1$, $\alpha_j = 0$ for $j \neq 1$ in (1). We compare the performance of multi-tier networks to this baseline, identifying conditions where multi-tier systems are advantageous. By adjusting tier-specific node properties (density, signal power, etc.), we regulate multi-tier system operation. Based on the physical properties of multi-tier networks, we assume decreasing signal power across tiers, that is,

$$\mu_1 \leq \mu_2 \leq \cdots \leq \mu_J. \tag{6}$$

For comparability, systems must be normalized. In the single-tier case, the average cell volume is $1/\lambda_1$ with bandwidth $w = \lambda_0/\lambda_1$. Extending to multiple tiers, we introduce $\lambda_i \geq 0$ for $i \geq 2$, keeping $\lambda_1$ fixed and increasing $\lambda_0$, leading to a total density $\lambda = \sum_{i=1}^{J} \lambda_i$ and preserving $w = \lambda_0/\lambda$. Denser BS and user deployment in multi-tier networks intensifies interference, yet lower-tier BSs operate with weaker signals according to (6), mitigating interference.

## 2.2  The Coverage Probability for Given Service Activity $\bar{q}$

A transmission event in a cell equipped with an $i$-th tier BS at $x$ is considered successful whenever the signal-to-noise ratio exceeds $t_i$, where $t_i$ is a predefined critical threshold specific to tier $i$. The probability of successful transmission along the link $(0, x)$ is reduced by the total interference $\mathcal{I}^{(i)}(x)$ from active base stations of any tier in surrounding cells, the positions of which we refer to in writing $\Phi_x$. The coverage probability is obtained from the probabilities of successful transmission,

$$V_{t_i}^{(i)}(\bar{q}; x, \Phi_x) = \mathbb{P}(\text{SIR}^{(i)}(x) > t_i)|\Phi_x),$$

$$V_{t_i}^{(i)}(\bar{q}; x) = \mathbb{E}[\mathbb{P}(\text{SIR}^{(i)}(x) > t_i|\Phi_x)], \text{ and}$$

$$V_{t_i}^{(i)}(\bar{q}) = \mathbb{E}[\mathbb{P}(\text{SIR}^{(i)}(X) > t_i)],$$

by averaging over the interferer Poisson locations $\Phi_x$ and the random location $X$ in (2) of the current transmitter link connection, successively in each component $i$. Since $t \mapsto 1 - V_t(\bar{q})$ is a probability distribution function, we let $\mathrm{SIR}^{(i)}$ be a random variable such that $\mathbb{P}(\mathrm{SIR}^{(i)} > t) = V_{t_i}^{(i)}(\bar{q})$. For $\bar{t} = (t_1, \ldots, t_J)$, the total coverage probability $V_{\bar{t}}(\bar{q})$, that is, the probability that a user at the origin successfully receives a signal sent from a BS of any tier, is

$$V_{\bar{t}}(\bar{q}) = \sum_{i=1}^{J} \alpha_i \, \mathbb{P}(\mathrm{SIR}^{(i)} > t_i) = \sum_{i=1}^{J} \alpha_i V_{t_i}^{(i)}(\bar{q}), \tag{7}$$

with weights given by the association probabilities $\{\alpha_i\}$ introduced in (1).

**Lemma 1.** *Take $\sigma^2 = 0$. With $\delta = \beta/d > 1$, the conditional coverage probability of a tier $i$ BS is*

$$V_{t_i}^{(i)}(\bar{q}) = \frac{q_i}{1 + w \sum_{j=1}^{J} \alpha_j q_j H_\delta(t_i \mu_i / \mu_j)}, \quad H_\delta(t) = \int_1^\infty \frac{1}{1 + u^\delta/t} \, du, \quad t \geq 0.$$

*Proof.* A signal link connection between a UE at the origin and a tier-$i$ BS at $x \in \mathbb{R}^d$ with interferer nodes in $\Phi_x$ is covered with probability

$$V_{t_i}(\bar{q}; x, \Phi_x) = q_i \, \mathbb{E}[e^{-\mu_i t_i \mathcal{I}^{(i)}(x)/a_\beta(x)} | \Phi_x].$$

Let $\theta_i(x) = \mu_i t_i / a_\beta(x)$. Then, taking expectations in (5) first with respect to $Q_j$ and then over $S_j$, for $j = 1, \ldots, J$, and using the mutual independence of the Poisson point processes $\Phi^j$, it follows that

$$\mathbb{E}\left[e^{-\theta_i(x)\mathcal{I}^{(i)}(x)} \,|\, \Phi_x\right] = \exp\left\{ \sum_{j=1}^{J} \int_{\mathbb{R}^d} \ln\left(1 - \frac{q_j \theta_i(x) a_\beta(y)}{\mu_j + \theta_i(x) a_\beta(y)}\right) 1_{\{|y| > |x|\}} \Phi^j(dy)\right\},$$

where the summation over $j$ represents the contribution of additional interference from tier-$j$ BSs. Next, we evaluate the expected values of the exponential Poisson integrals to obtain

$$\mathbb{E}\left[e^{-\theta_i(x)\mathcal{I}^{(i)}(x)}\right] = \exp\left\{ -\sum_{j=1}^{J} \int_{|y| > |x|} \frac{\lambda_0 \alpha_j q_j \theta_i(x) a_\beta(y)}{\mu_j + \theta_i(x) a_\beta(y)} \, dy, \right\}.$$

and therefore

$$V_{t_i}^{(i)}(\bar{q}; x) = q_i \exp\left\{ -\int_{|y| > |x|} \sum_{j=1}^{J} \frac{\lambda_0 \alpha_j q_j \mu_i t_i a_\beta(y)}{\mu_j + \mu_i t_i a_\beta(y)/a_\beta(x)} \, dy \right\}.$$

By a change of variables using $v = |x/\ell_0|^d$ and $u = |y/\ell_0|^d$, the $dy$-integral simplifies since $a_\beta$ is a radial function. Then, averaging the distribution in (2),

$$V_{t_i}^{(i)}(\bar{q}) = \mathbb{E}[V_{t_i}^{(i)}(\bar{q}, X)] = \int_0^\infty V_{t_i}^{(i)}(\bar{q}; v^{1/d}\ell_0) \, e^{-v} \, dv,$$

yields the result stated in the lemma. $\square$

Next, we say that the multi-tier system has *biased load balancing* [4], whenever the threshold parameters $\bar{t}$, satisfy

$$t_i\mu_i = t_1\mu_1, \quad i = 2.\ldots, J. \tag{8}$$

The idea of biased load balancing is that if the SIR thresholds decrease sufficiently fast and follow the signal power order, then the multi-tier system is expected to provide larger coverage than the single-tier system. This assumption aligns with BS characteristics, where macro-BSs tolerate higher thresholds than micro-BSs. Physically, the setting in (8) effectively expands the coverage radius of the smaller nodes by reducing the SIR threshold. To assess the impact of load balancing, let

$$U_{t_i}^{(i)}(\bar{q}) = V_{t_i}^{(i)}(\bar{q})/q_i, \tag{9}$$

denote the conditional probability of coverage in tier $i$, given that the BS is busy. Using Lemma 1, we immediately have the following observation. Under biased load balancing, the probabilities $U_{\bar{t}}^{(i)}(\bar{q})$ in (9) are the same across all tiers, so

$$U_{t_i}^{(i)}(\bar{q}) = U_{t_1}^{(1)}(\bar{q}), \quad i = 2,\ldots, J. \tag{10}$$

It remains to associate $\bar{q}$ with the buffer distribution, using (4). We will next see that if the system has biased load balancing, then we can find $q = q^*(p, \bar{t})$ as a function of $p$ and $\bar{t}$ such that (4) holds.

## 3   Summary of Results

To enable signal transmission in the cellular network equipped with buffers of maximal size $K$, the input data rate $p$ must be balanced by the system response parameters $\bar{q} = (q_1, \ldots, q_J)$, to avoid that the nodes become overloaded or under-utilized. In [6, Lemma 4, Theorem 1], the stationary distribution of the buffer size is derived in a truncated geometric form for $J = 1$. For $K < \infty$, the empty buffer probability $\mathbb{P}(B = 0)$ is such that Eq. (4) has the equivalent form

$$q = 1 - \frac{U(q) - p}{U(q) - p\,b(q)^K}, \qquad b(q) = \frac{p(1 - U(q))}{(1 - p)U(q)}, \qquad U(q) = \frac{V_t(q)}{q}, \tag{11}$$

having a unique, minimal solution $q^*$, such that $p < q^* \leq 1$. This approach also applies to the multi-tier situation, where we seek $\bar{q}$ such that

$$q_i = 1 - \frac{U^{(i)}(q_i) - p}{U^{(i)}(q_i) - p\,b(q_i)^K}, \qquad 1 = 1, \ldots, J.$$

However, assuming biased load balancing, the relation (10) shows that these $J$ equations are identical. Hence, there is a single solution $q^* = q_1^* = \cdots = q_J^*$.

For the case of a multi-tier no-loss system with unlimited buffers, $K = \infty$, the system of balance equations is $V_{t_i}^{(i)}(\bar{q}) = p$, for each $i$. Applying biased load balancing to this situation, there is again a single solution $q^*$, now assuming that the net arrival rate $p$ is sufficiently small.

**Proposition 1.** *Assume that the multi-tier network satisfies biased load-balancing (8) with $t_1 = t$.*

*$K < \infty$. The balance equation (11) with*

$$U(q) = \frac{1}{1 + C_t q}, \qquad C_t = w \sum_{j=1}^{J} \alpha_j H_\delta(\mu_1 t / \mu_j),$$

*has a unique minimal solution $q^* = q(p, t)$. The probability that the buffer is empty is $P(B^i = 0) = (1 - q(p, t))/(1 - p)$ in each tier, and the total throughput in (7) is $V_t(q(p, t)) = q(p, t)/(1 + C_t\, q(p, t))$.*

*$K = \infty$. For each $p \leq 1/(1 + C_t)$, the cells of the no-loss system are active with probability $q^* = q(p, t)$, where*

$$p < q(p, t) = \frac{p}{1 - C_t\, p} \leq 1.$$

As an example of how to use Proposition 1, we find for the case $K = 1$ that the minimal solution $q^* = q(p, t)$ of the second-order equation

$$\frac{q}{1 + C_t q} = \frac{p(1 - q)}{1 - p}, \qquad C_t = w \sum_{j=1}^{J} \alpha_j H_\delta(\mu_1 t / \mu_j),$$

is the probability of BS activity

$$q(p, t) = \frac{1}{2pC_t}\left(\sqrt{(1 - pC_t)^2 + 4p^2 C_t} - (1 - pC_t)\right).$$

**Shannon Capacity.** We now extend the performance analysis to evaluate the channel capacity for the multi-tier system. The Shannon capacity is

$$\mathrm{Cap}_K^{(J)}(p, w) = w \sum_{i=1}^{J} \alpha_i \mathbb{E}[\log_2(1 + \mathrm{SIR}^{(i)})]. \tag{12}$$

In [7], for single-tier networks, we tentatively proposed the property of the network being *within Shannon capacity*, defined as $\mathrm{Cap}_K^{(J)}(p, w) \geq pw$, as a performance measure. Under the same assumptions as in Proposition 1, using a log-integral representation in (12), we have

$$\mathrm{Cap}_K^{(J)}(p, w) = \mathbb{E}\left[ \sum_{i=1}^{J} \alpha_i \int_0^{\mathrm{SIR}^{(i)}} \frac{dt_i}{1 + t_i} \right]$$

$$= \frac{w}{\ln 2} \sum_{i=1}^{J} \alpha_i \int_0^{\infty} \frac{V_{t_i}^{(i)}(q(p, t))}{1 + t_i}\, dt_i = \frac{w}{\ln 2} \int_0^{\infty} \frac{V_t(q(p, t))}{1 + t}\, dt, \tag{13}$$

which can be used to investigate the property mentioned, $\mathrm{Cap}_K^{(J)}(p, w) \geq pw$.

**Reference case: $K = 0$.** For the pure loss network $K = 0$, without BS buffers, the activity level matches the input rate. Thus, $q_i = p$ for all $i$, and the total coverage probability is given by $V_{\bar{t}}(p)$. This allows us to use a condition similar but weaker than (8), as stated in the following proposition.

**Proposition 2.** *For the case $K = 0$, the Shannon capacity is*

$$\mathrm{Cap}_0^{(J)}(p, w) = \frac{1}{\ln 2} \int_0^\infty \frac{pw}{1 + wp \sum_{j=1}^J \alpha_j H_\delta(t\mu_1/\mu_j)} \sum_{i=1}^J \frac{\alpha_i}{\mu_1/\mu_i + t} \, dt.$$

*If $\mu_i t_i \leq \mu_{i-1} t_{i-1}$, $i = 2, \ldots, J$, then the multi-tier system performs better than the single-tier case, that is,*

$$V_{\bar{t}}(p) \geq p/(1 + wpH_\delta(t_1\mu_1)),$$

*where the r.h.s. is obtained by inserting $J = 1$ in (7), reflecting the single-tier. Moreover,*

$$\mathrm{Cap}_0^{(J)}(p, w) \geq \frac{1}{\ln 2} \int_0^\infty \frac{pw}{1 + wpH_\delta(t_1\mu_1)} \frac{1}{1 + t} \, dt.$$

**No-loss case: $K = \infty$.** In Proposition 1, the solution $q(p, t)$ for the no-loss system is obtained for a fixed $t$ and small enough $p$, such that $p \leq 1/(1 + C_t)$. Equivalently, for fixed $p$ there is a well-defined solution $q(p, t) \leq 1$ of the balance equation $V_t^{(i)}(q) = p$ for every threshold $t \leq T_{\max}^{(J)}(p, w)$, where

$$T_{\max}^{(J)}(p, w) = \sup\left\{ t > 0 : \sum_{j=1}^J \alpha_j H_\delta(\mu_1 t/\mu_j) \leq \frac{1}{w}\left(\frac{1}{p} - 1\right) \right\}.$$

This constraint is only required in the infinite buffer case; when the buffer is finite, the stationary solution exists for any $t$. Compared to the case $J = 1$ of a single tier, recalling that $\mu_1 \leq \mu_j$ for all $j$, we see that $T_{\max}^{(1)}(p, w) \leq T_{\max}^{(J)}(p, w)$. Hence, the average rate at which the network can be reliably operated is larger for the multi-tier system. We can now show that this property carries over to the Shannon capacity in (13). Given $J$, for fixed $p$ and $w$, the solution $q(p, t)$ in Proposition 1 is a function of $t$, such that

$$\mathbb{P}(\mathrm{SIR}^{(1)} > t) = V_t^{(1)}(q(p, t)) = p,$$

for every $t \leq T_{\max}^{(J)}(p, w)$. Thus,

$$\mathrm{Cap}_\infty^{(J)}(p, w) = w \int_0^{T_{\max}^{(J)}(p,w)} \frac{p}{1 + t} \, dt = pw \log_2(1 + T_{\max}^{(J)}(p, w)).$$

Consequently, the multi-tier no-loss network is within Shannon capacity whenever $T_{\max}^{(J)}(p, w) > 1$, that is,

$$\frac{pw}{1 - p} \leq \frac{1}{\sum_{j=1}^J \alpha_j H_\delta(\mu_1/\mu_j)}.$$

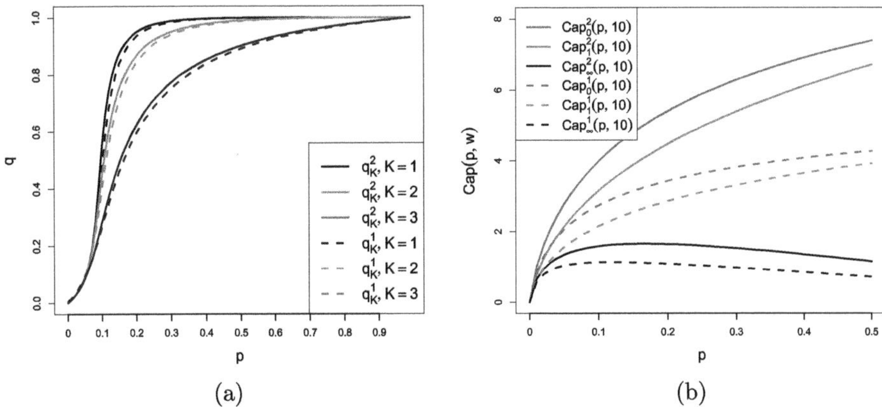

**Fig. 1.** a) Busy link probability and b) Shannon capacity for $J = 1, 2$ for $w = 10, c = 0.5, \alpha_2 = 0.8$, for different buffer sizes.

More generally, given $J$ and $K \geq 2$, we return to the corresponding nonlinear equation (11) and use the result stated in Proposition 1 to solve for the unknown $q^*$. The relevant capacity $\mathrm{Cap}_K^{(J)}(p, w)$ is then obtained from (13). Examples of such solution functions $q^* = q_K^J$, for $J = 2$ and $J = 1$ with selected values of $K$ are shown in Fig. 1(a). Figure 1(b) shows the corresponding Shannon capacity $\mathrm{Cap}_K^J(p, w)$ for $J = 2$ with $K = 0, 1, \infty$, as a function of $p$, compared to the single-tier case. Additionally, a special case of the pure-loss system is demonstrated in Fig. 2, where the capacity is shown for various values of $\alpha_1, \alpha_2$ and $t_1, t_2$. The results demonstrate that, for all considered buffer sizes $K$, the capacity is consistently higher in the two-tier system.

## 4 Multi-tier Network with Different Path Loss

From the results presented in previous sections, we have seen that the multi-tier network with varying signal power performs better compared to the single-tier system under balancing of the load. The gain in coverage and capacity is limited because of the increase in the interference field. A further option to mitigate interference is to adopt a tier-specific path loss model. In this direction, we investigate a variation of the multi-tier model restricted to pure loss, and consider a network where transmitting nodes of different tiers have different path loss behavior.

**Lemma 2.** *In the multi-tier network with tier-dependent path loss exponents $\beta_i$ and associated attenuation functions*

$$a_i(x) = |x/\ell_0|^{-\beta_i}, \quad i = 1, \ldots J,$$

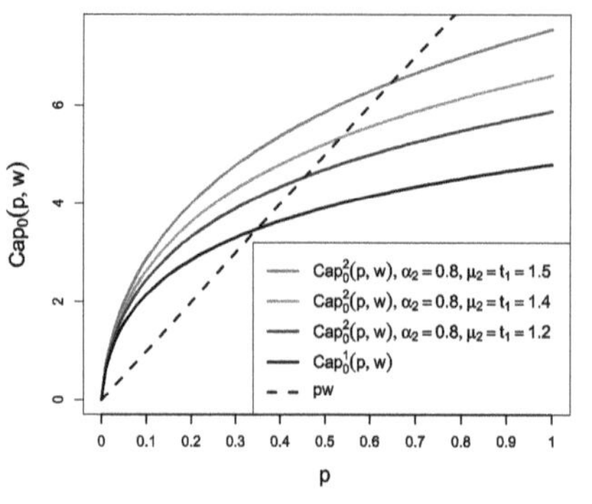

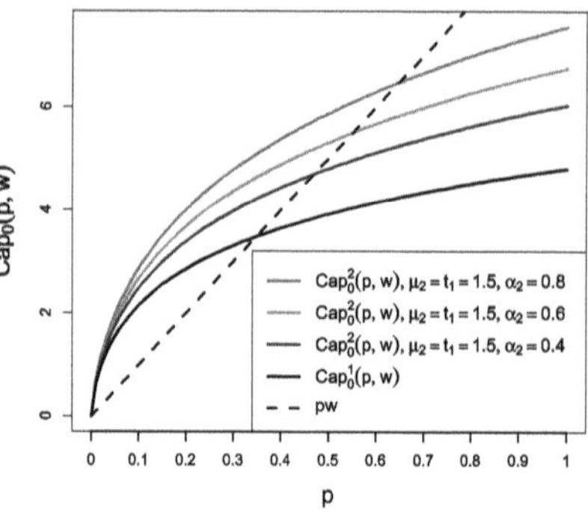

(a) Shannon capacity as a function of $p$ for different $\mu_2$ and $t_1$, with $\mu_1 = t_2 = 1$ and $\alpha_2 = 0.8$, pure loss.

(b) Shannon capacity as a function of $p$ for different $\alpha_2$, with $\mu_1 = t_2 = 1$ and $\mu_2 = t_1 = 1.5$, pure loss.

**Fig. 2.** Shannon capacities $\mathrm{Cap}_0^{(1)}(p,w)$ and $\mathrm{Cap}_0^{(2)}(p,w)$.

*the total coverage probability for the pure-loss case is given by*

$$V_{\bar{t}}(p) = \sum_{i=1}^{J} p\alpha_i \int_0^\infty \exp\left\{ -\int_v^\infty \Sigma_{uv}^{(i)}(p)\,du \right\} e^{-v}\,dv, \qquad (14)$$

*where*

$$\Sigma_{uv}^{(i)}(p) = \sum_{j=1}^{J} \frac{\alpha_j pw}{1 + u^{\delta_i}/(v^{\delta_j} t_i \mu_i/\mu_j)}, \qquad \delta_i = \beta_i/d, \quad i = 1,\dots,J.$$

*Proof.* The proof relies on the same arguments as in Lemma 1. The interference is generated again from tiers of all classes, but now signals from each tier have a different attenuation $a_i(\cdot), i = 1,\dots,J$. As in (5), the interference affecting a tier-$i$ BS, is now

$$\mathcal{I}^{(i)}(x) = \sum_{j=1}^{J} \sum_{y\in\Phi^j\setminus x} a_i(y)S_y^{(j)}Q_y^{(j)}.$$

Letting $\theta_i(\cdot) = \mu_i t_i/a_i(\cdot)$, by analogy with the previous analysis, the logarithmic Laplace transform of the interference for tier $i$ is

$$-\ln\mathbb{E}[e^{-\theta_i(x)\mathcal{I}^{(i)}(x)}] = \sum_{j=1}^{J} \int_{|y|>|x|} \frac{\lambda_0\alpha_j p\,\theta_i(x)a_j(y)}{\mu_j + \theta_i(x)a_j(y)}\,dy$$

$$= \sum_{j=1}^{J} \int_{|y|>|x|} \frac{\lambda_0 p\,\alpha_j}{1 + (\mu_j/\mu_i t_i)a_i(x)/a_j(y)}\,dy.$$

Employing once more the variable change $v = |x/\ell_0|^d$, $u = |y/\ell_0|^d$, we get

$$- \ln \mathbb{E}[e^{-\theta_i \mathcal{I}^{(i)}}] = \alpha_i pw\, v \int_1^\infty \frac{1}{1 + u^{\delta_i}/t_i}\, du$$

$$+ w \sum_{k=1,k\neq i}^{J} \alpha_k p \int_v^\infty \frac{1}{1 + (u^{\delta_i}/v^{\delta_k})/(t_i\mu_i/\mu_k)}\, du.$$

Now putting things together as in the proof of Lemma 1, we get the expression (14), and the proof is complete.

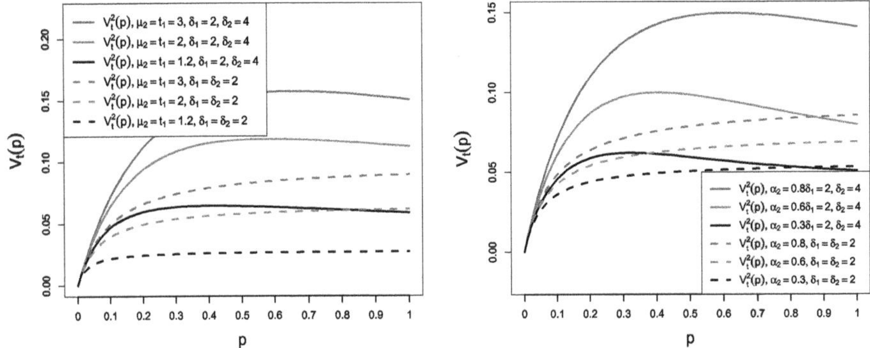

(a) Pure-loss coverage probability as a function of $p$ for different values of $\delta_1, \delta_2$, for $t_1 = 3$, $t_2 = 1$, $\mu_1 = 1$, $\mu_2 = 3$, $\alpha_2 = 0.8$.

(b) Pure-loss coverage probability as a function of $p$ for different values of $\delta_1, \delta_2$, and $\alpha_2$, for $\mu_1 = t_2 = 1$ and $\mu_2 = t_1 = 4$.

**Fig. 3.** Comparison of pure-loss coverage probability for the different path loss exponents for $w = 10$, $t_1 = 2$, $t_2 = 1$.

We note that in the special case $J = 2$ of the two-tier model the coverage probability can be further simplified. Since the expression (14) depends only on the ratios of $\mu_i, i = 1, \ldots, J$, let $\hat{\mu} = \mu_2/\mu_1$, to allow for parameter reduction. Then the coverage probability is obtained as

$$V_{\bar{t}}(p) = p\alpha_1 \int_0^\infty \exp\left\{ - w\left(\alpha_1 pv \int_1^\infty \frac{du}{1 + u^{\delta_1}/t_1}\right.\right.$$

$$\left.\left. + \alpha_2 p \int_v^\infty \frac{du}{1 + (u^{\delta_1}/v^{\delta_2})/(t_1/\hat{\mu})}\right)\right\} e^{-v}\, dv$$

$$+ p\alpha_2 \int_0^\infty \exp\left\{ - w\left(\alpha_1 p \int_v^\infty \frac{du}{1 + (u^{\delta_2}/v^{\delta_1})/(t_2\hat{\mu})}\right.\right.$$

$$\left.\left. + v\alpha_2 p \int_1^\infty \frac{du}{1 + u^{\delta_2}/t_2}\right)\right\} e^{-v}\, dv. \tag{15}$$

Unlike the previous case $\delta_i = \delta$ (i.e. $\beta_i = \beta$), the integral expressions (14) or (15) do not simplify further. Using a numerical computation, examples of the coverage probabilities $V_{\bar{t}}(p)$ for $J = 2$ in (15) are shown in Fig. 3. For comparison, we present the numerical coverage probability for the two cases of the same and different path loss exponents, plots (a) and (b), respectively. Note that we choose path loss exponents such that $\beta_1 < \beta_2$. Physically, this means that the nodes with weaker signal power transmit information over shorter distances compared to the stronger nodes. We see that the deployment of additional stations results in a performance gain under both identical and tier-specific path loss. However, the increase in coverage is noticeably larger when using tier-specific path-loss exponents. Hence, the model with different path loss exponents is more advantageous, with smaller cells having higher path loss values. A similar discussion is presented in [4]. The authors show that, for a special case of the pure-loss system where all nodes are active, dense deployment of the smaller-tier nodes results in a performance gain only when the path-loss exponents are tier-specific.

## 5   Simulation Results

We now proceed to validate the proposed model using stochastic simulation, focusing on a multi-tier network. The simulation employs a discrete-time stochastic framework over an $R \times R$ area with mobile users and BSs across tiers, whose locations follow independent Poisson point processes with intensities $\lambda_0$ (users) and $\lambda_i$ (BSs of tier $i$). With fixed node positions, we simulate network behavior for up to $N$ slots and estimate performance metrics via Monte Carlo sampling. Full simulation details and single-tier validation are provided in [6]. The most relevant performance metrics for the experiment setting appear to be coverage probability $V_t(q)$ and activity probability $q(p)$ in a multi-tier network. The goal is to compute the estimates of $V_t(q)$ and $q(p)$ and compare them to the metrics derived analytically. The experiments are performed for the cases of pure loss ($K = 0$) and limited loss ($0 < K < \infty$). In the first experiment, we simulate a two-tier pure-loss system ($K = 0$) with $p = 0.5$, $N = 10000$, $t_1 = 5$, and $\mu_1 = 1$. We estimate the average coverage probability $\hat{V}t(p)$ as a function of $\mu_2$, along with tier-specific coverage $\hat{V}t_1^{(1)}(p)$ and $\hat{V}_{t_2}^{(2)}(p)$, for varying $\beta$. Results for $\delta_1 = \delta_2 = 2$ and $\delta_1 = \delta_2 = 4$ are shown in Fig. 4(a). The second experiment examines the limited-loss case ($0 < K < \infty$), estimating the activity probability $q$ for different buffer sizes. With $\delta = 2$, $t_1 = t_2 = 2$, $\mu_1 = 1$, and $\mu_2 = 2$, we compute $q$ as a function of $p$, shown in Fig. 4(b). We see that the estimates obtained from the simulation are relatively close to the theoretical values, which validates the accuracy of the proposed model. We also note that removing weights from the respective tier-specific probability, i.e., putting

$$\hat{V}_{t_i}^{(i)}(p) = \hat{V}_{t_i}^{(i)}(p)/\alpha_i, i = 1, 2$$

will result in $\hat{V}_{t_1}^{(1)}(p) \approx \hat{V}_{t_2}^{(2)}(p)$, which is consistent with the result (10). Note that in most plots, the estimates for the 2nd tier are off more than those for the

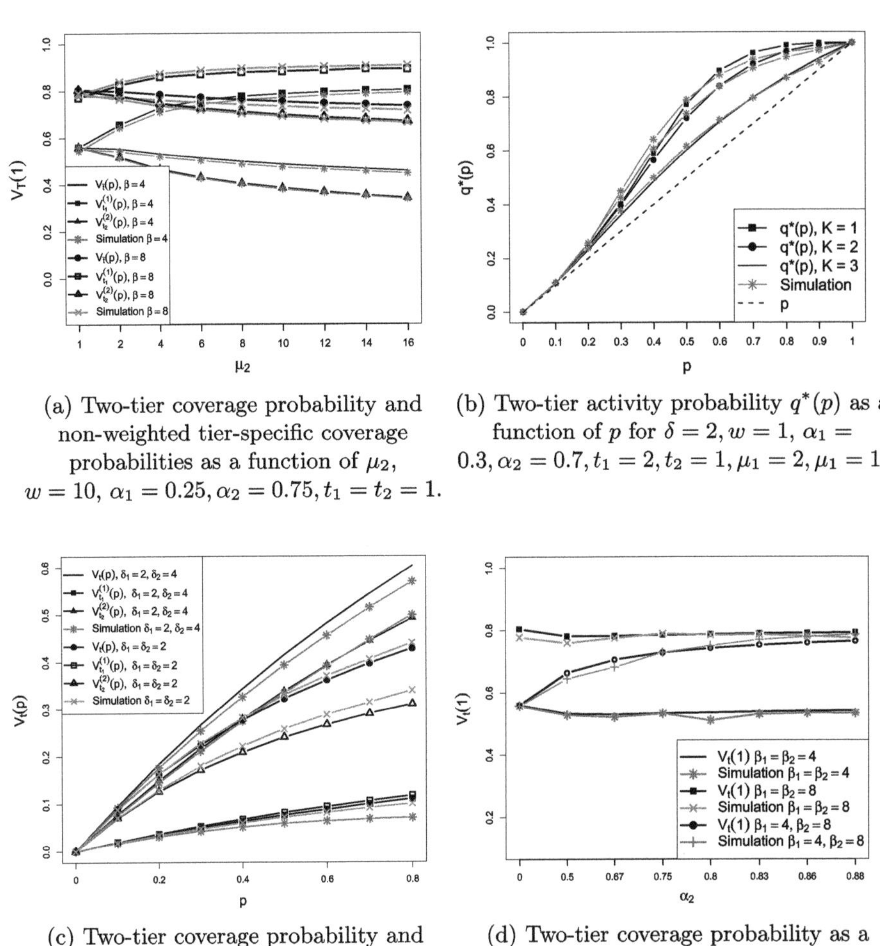

(a) Two-tier coverage probability and non-weighted tier-specific coverage probabilities as a function of $\mu_2$, $w = 10$, $\alpha_1 = 0.25, \alpha_2 = 0.75, t_1 = t_2 = 1$.

(b) Two-tier activity probability $q^*(p)$ as a function of $p$ for $\delta = 2, w = 1, \alpha_1 = 0.3, \alpha_2 = 0.7, t_1 = 2, t_2 = 1, \mu_1 = 2, \mu_1 = 1$.

(c) Two-tier coverage probability and non-weighted tier-specific coverage probabilities as a function of $p$, $w = 10$, $\alpha_1 = 0.2, \alpha_2 = 0.8, t_1 = 2, t_2 = 1, \mu_2/\mu_1 = 8$.

(d) Two-tier coverage probability as a function of $\alpha_2$, $w = 10$, $t_1 = t_2 = 1$, $\hat{\mu} = 4$.

**Fig. 4.** Two-tier coverage probability as a function of $\alpha_2$ and simulation estimates for no noise system where $t = 1$ and $\hat{\mu} = 4$.

1st one and overall coverage. This difference is caused by the choice of network parameters, where we set $\alpha_1 = 0.2$ and $\alpha_2 = 0.7$. As a result, there are on average fewer BSs of the 1st tier in the network, hence averaging over the cells gives a less accurate result. The last two experiments validate the findings from Sect. 5, comparing a multi-tier network with tier-specific path loss exponents to the reference case of the same path loss. We set $\delta_1 = 2$ and $\delta_2 = 4$ (corresponding to $\beta_1 = 4$, $\beta_2 = 8$ for $d = 2$), expecting the lower exponent in the second tier to reduce interference and improve coverage. Figure 4 (b) confirms this: both analytical and simulation results show higher coverage for $\delta_1 = 2, \delta_2 = 4$ than for the case when $\delta_1 = \delta_2 = 4$. This suggests that using different path loss exponents, especially in dense deployments of low-power nodes, can improve overall network performance and reduce energy costs by deploying less expensive small BSs.

# References


1. Andrews, J., Baccelli, F., Ganti, R.K.: A tractable approach to coverage and rate in cellular networks. IEEE Trans. Commun. **59**, 3122–3134 (2011). https://doi.org/10.1109/TCOMM.2011.100411.100541
2. Dhillon, H.S., Ganti, R.K., Baccelli, F., Andrews, J.G.: Modeling and analysis of K-tier downlink heterogeneous cellular networks. IEEE J. Sel. Areas Commun. **30**(3), 550–560 (2012). https://doi.org/10.1109/JSAC.2012.120405
3. Gupta, A.K., Dhillon, H.S., Vishwanath, S., Andrews, J.G.: Downlink multi-antenna heterogeneous cellular network with load balancing. IEEE Trans. Commun. **62**(11), 4052–4067 (2014). https://doi.org/10.1109/TCOMM.2014.2350504
4. Jo, H.S., Sang, Y.J., Xia, P., Andrews, J.G.: Heterogeneous cellular networks with flexible cell association: a comprehensive downlink SINR analysis. IEEE Trans. Wireless Commun. **11**(10), 3484–3495 (2012). https://doi.org/10.1109/TWC.2012.081612.111361
5. Dhillon, H.S., Novlan, T.D., Andrews, J.G.: Coverage probability of uplink cellular networks. In: 2012 IEEE Global Communications Conference (GLOBECOM), pp. 2179–2184 (2012)
6. Kaj, I., Morozova, T.: Retransmission performance in a stochastic geometric wireless communication model. Perform. Eval. **165**, 102428 (2024). https://doi.org/10.1016/j.peva.2024.102428
7. Kaj, I., Morozova, T.: Channel capacity and performance analysis of a buffered cellular network. TechRxiv (2024). https://doi.org/10.36227/techrxiv.172565778.87538590/v1
8. Mahbub, M., Alharbi, A.G., Barua, B.: Maximization of the User association of a low-power tier deploying biased user association scheme in 5G multi-tier heterogeneous network. In: 2021 IEEE 12th Annual Information Technology, Electronics and Mobile Communication Conference (IEMCON), Vancouver, BC, Canada, pp. 0928–0933 (2021)
9. Naghshin, V., Reed, M.C.: On capacity and association area characterization in small cell-based multi-tier networks. IEEE Wirel. Commun. Lett. **4**(5), 505–508 (2015)

# Performance-Aware Microservices Architecture Live Planning and Scaling


A. Capizzi[1], G. Mancini[2], M. Scarpa[3(✉)], and S. Distefano[3]

[1] Cleafy S.p.A, Milan, Italy
`antonio.capizzi@cleafy.com`
[2] Sobereye Inc., Menlo Park, USA
[3] University of Messina, Messina, Italy
`{mscarpa,sdistefano}@unime.it`



**Abstract.** Microservices architecture (MSA), with containerization and Cloud provisioning, is now a standard for software deployment, driven by automation and DevOps practices. While improving modularity, flexibility and maintainability, MSA presents challenges in non-functional quality aspects, including security, performance, and reliability. This paper addresses the open problem of adaptive, real-time capacity planning for performance-driven scaling of microservices, focusing on the MSA API gateway pattern. A queuing-network-based methodology is developed to estimate the microservice replicas per workload by explicitly modeling infrastructure and interactions. This approach enables accurate, flexible infrastructure scaling capacity planning and design-time analysis of system properties. A 3-step methodology including benchmaring, modeling, and deployment is proposed and applied to a real-world case study on an API gateway MSA to demonstrate its effectiveness.

**Keywords:** Capacity Planning · Microservices · API Gateway · Orchestrator · Scaling · Response Time · Queueing Networks · Endpoint


## 1 Introduction

The software design, implementation and deployment have nowadays mainly transitioned to microservices architecture (MSA) that, integrated with containerization and cloud provisioning, establishes a mainstream solution, a de-facto standard. This is driven by the automation provided by software-defined data centers, Infrastructure-as-Code, and agile/DevOps pipelines and toolchains, enabling faster development, improved scalability, and flexibility. The modularity of MSA also eases updating and maintaining complex applications, with cloud and virtualization technologies further enhancing efficiency and consistency across deployments.

However, MSA introduces challenges, particularly in ensuring non-functional quality aspects such as security, performance, reliability, maintainability, availability, resilience, and cost-effectiveness. The distributed nature of MSA complicates security and requires robust AAA mechanisms. Performance is affected by



L. Carnevali and J. Doncel (Eds.): EPEW 2025, LNCS 15657, pp. 101–114, 2026.
https://doi.org/10.1007/978-3-032-16345-5_8

workload fluctuations and network latency, requiring efficient load balancing and scaling policies. Reliability, maintainability, availability, and resilience are critical due to potential cascading failures asking for fault tolerance and redundancy. Cost-effectiveness is also a concern, given the potential for increased overhead, such as overprovisioning.

Solutions such as container isolation, API gateways and MSA orchestrators address some issues. Container isolation mainly improves security while API gateways and orchestrators enhance both security (integrating AAA) and performance (through load balancing and autoscaling). MSA itself promotes reuse, simplifies modifications, and contains faults, improving reliability, maintainability, availability, and resilience. However, open problems still persist, notably achieving effective scaling through adaptive, real-time capacity planning for performance-driven autoscaling.

In this context, model-driven technique may be quite effective in analyzing MSA-based software system performance and predicting workloads and behaviours accordingly. However, current black-box modeling approaches lack the visibility required for efficient resource allocation. To such a purpose, this work proposes a white-box, queuing-network-based capacity planning methodology to implement performance-aware MSA live planning and scaling (PAM-SALPS). By explicitly modeling infrastructure and microservice interactions, the proposed methodology estimates microservice replicas per workload based on performance and cost tradeoffs, minimizing resource allocation. This technique is able to implement accurate, flexible infrastructure capacity planning while enabling design-time analysis of system properties, including workload fluctuations.

The main contributions of this work are: (1) a comprehensive microservice model including Web and application server adopting endpoint-level modeling for finer granularity; (2) MSA modeling integrating the API gateway, the infrastructure orchestrator, and all microservices and related endpoints; (3) a methodology for capacity planning of API gateway-based MSA; (4) a performance-aware scaling heuristics; and (5) a real-world case study and validation. Details in the remainder of this paper, structured as follows. Section 2 provides preliminary concepts, Sect. 3 details the proposed approach, Sect. 4 presents a case study involving a real API gateway-based MSA. Section 5 closes the paper with remarks and future work.

## 2  Preliminary Concepts

A microservices architecture (MSA) [7] decomposes large applications into small autonomous services performing specific tasks within bounded contexts. MSA offers advantages including fine-grained scalability, maintainability, agility, reduced codebase, reuse, focused teams, data isolation, and technology diversity. Microservices typically expose RESTful API or *endpoints* [18] through Web servers [25]. Web servers generally include a request queue, a thread pool for processing, and a connection manager for concurrent connections. Deploying Web

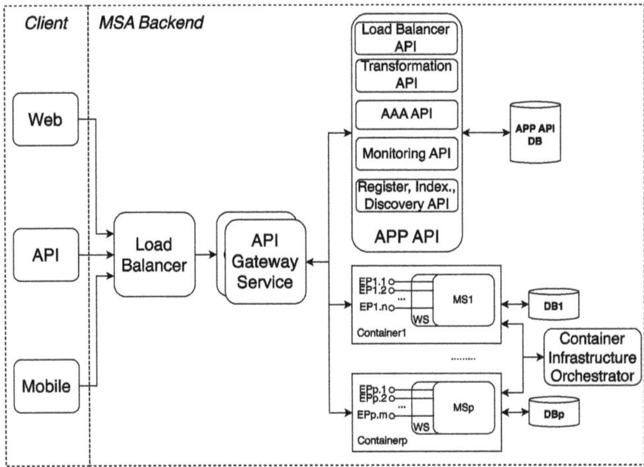


**Fig. 1.** API Gateway MSA pattern

servers with each microservice improves API/endpoint scalability and management, through independent versioning and deployment.

Microservices provide from one API or endpoint to many, which in an MSA, altogether, may be quite a large number to be properly indexed, discovered, and managed. For their management, MSA thus often recurs to the API Gateway, a single entry point acting as a reverse proxy with cross-cutting functionalities (routing, TLS termination, AAA, load balancing), and an orchestrator for the microservice deployment and scaling. The API Gateway pattern [26], shown in Fig. 1 simplifies client interfaces, centralizes security, improves scalability and performance, and streamlines microservice implementation. Microservices communicate internally through synchronous (REST, RPC) or asynchronous (message queues such as Apache Kafka [1]) calls and are designed for independent scalability and replaceability. Examples include Netflix and its Zuul framework[1]. Summarizing, an API gateway MSA managing $m > 1$ microservices, with $e_i > 0$ endpoints for the $i^{th}$ microservice $ms_i$ ($i = 1, .., m$), exposes a number of endpoints $e > 0$ that may be $e \leq \sum_{i=1}^{m} e_i$ or even $e > \sum_{i=1}^{m} e_i$ including orchestrated microservice endpoints.

Fine-grained scaling [11] is a key MSA advantage, primarily implemented horizontally by adjusting microservice instances to workload fluctuations, enhancing maintainability, scalability, resilience, and cost-effectiveness. Horizontal scaling allows fine-grained resource adaptation to demand. Microservices are commonly deployed as containerized applications managed by orchestrators like Kubernetes [10] on virtual or physical machines. This involves virtualization layers and utilizes HTTP servers with thread pools or event loops for request handling.

Web server-based application modeling [23] is crucial for designing, implementing, analyzing, and optimizing Web services, including MSA. Modeling

---

[1] https://github.com/Netflix/zuul.

methods include queueing networks (QN) [8], simulation, Markov models, and black-box models. QN accurately model Web server architectures [5,17] by capturing queue behavior and client-server interactions. QN also suit MSA performance modeling [21,27], where microservices are interconnected Web servers.

Optimizing scaling policies in MSA is of primary importance. Tools like Kubernetes [10] provide threshold-based autoscaling. Research focuses on metrics [19] and thresholds [13] using machine learning [3]. Optimal autoscaling requires advanced frameworks. Prediction-driven autoscaling (ProScale [6], BHAS [20]), cost-efficient autoscalers ($\mu$Opt [12], PBScaler [24]), machine learning-based approaches (DCScaler [15]), reinforcement learning [14], concurrency management ($\mu$ConAdapter [16]), and model-driven autoscaling (ATOM [9,22]) exist. Distributed autoscaling [2] improves scalability.

Current MSA autoscaling often uses black-box models despite known MSA structures. White-box modeling offers advantages like lower data needs and higher accuracy for design and analysis. This paper proposes a white-box QN-based model for API gateway MSA autoscaling.

## 3   Proposed Approach

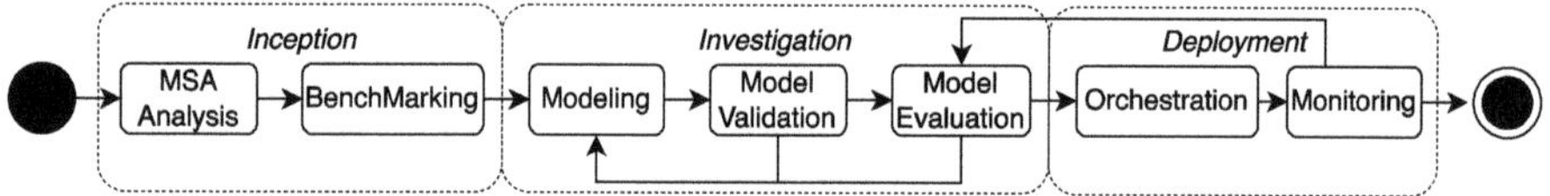


**Fig. 2.** Proposed approach workflow

This paper proposes a three-stage methodology for *performance-aware microservices architecture live planning and scaling (PAMSALPS)*, focusing on API gateway-based MSA. The PAMSALPS approach encompasses *inception, investigation*, and *deployment* stages to determine and validate optimal scaling configurations, as shown in Fig. 2. The proposed approach works at a finer-grained granularity than the other in the literature, considering endpoints instead of (higher-level) microservices. This means that the inception, investigation, and deployment stages focus on endpoints. Thereby, the analysis, benchmarking, modeling, evaluation, and deployment operate at the endpoint granularity, i.e. by analyzing, benchmarking, modeling, evaluating, deploying, and scaling endpoints.

### 3.1   Inception

The PAMSALPS inception phase is mostly tasked with an in-depth study of the MSA ecosystems. It involves an analysis and the comprehensive benchmarking of any single microservice and endpoints. It starts with the *MSA analysis*, aiming to identify the microservices architecture, the API Gateway, all the component microservices and, for each of them, the specific API and endpoints they expose. After this preliminary analysis, workload generation tools are exploited to benchmark each microservice and its publicly accessible endpoints to a spectrum of controlled and increasing request rates. This systematic workload testing aims to elicit the performance characteristics of the microservice endpoint under varying stress levels. The *benchmarking* process thus focuses on the collection of key performance indicators for each endpoint in the MSA. These include average request latency, percentile distributions of response times (e.g., demand and service times), throughput (requests per second), and resource utilization metrics (CPU, memory, network I/O) at each microservice instance and endpoints. The collected statistical data provide an empirical foundation for subsequent modeling, capturing the inherent performance profiles and potential bottlenecks within each microservice endpoint such as demand, service time and similar to the next modeling stage. The benchmarking strategy ensures coverage of all critical service interactions and API endpoints to provide a holistic view of system performance under load.

## 3.2   Investigation

The second PAMSALPS phase focuses on the investigation of the MSA performance. To such a purpose, a queuing network (QN) is exploited to represent the MSA in the *modeling* step. The topology of the QN model mirrors the interactions between the API gateway and the backend microservices, with each microservice endpoint represented as a service station in the QN model with associated queuing and service disciplines. The model parameters, such as service rates and arrival processes at each service station, are derived from the statistical data obtained during the benchmarking phase.

The modeling approach here adopted, differently from the literature ones, represents each microservice endpoint with a single station. The underlying assumption here introduced is to have at least one thread per endpoint. An endpoint station is multi-server, where the number of servers is the number of threads instantiated by the microservice. This latter is a reasonable assumption, because the main PAMSALPS goal is to determine the appropriate number of resources for managing critical workloads stressing the system. It is worth noting that the thread pool is slightly overestimated in light workloads, but when the system workload increases the PAMSALPS approximation can be effective. Requests to different endpoints are implemented with different user classes. Incoming requests are forwarded to the appropriate microservice on a statistical basis through a QN router, modeling the API gateway routing. Furthermore, the API gateway MSA is modeled in PAMSALPS by associating with each endpoint exposed by the API gateway a specific sink and a path from the API gateway router to the sink. This may involve multiple microservice endpoints and thus related stations in the corresponding QN model, even shared by the API gateway endpoints.

Modeling each microservice with multiple stations, specifically one per endpoint, addresses the non-parallel processing of HTTP requests within a web server. Representing an entire microservice as a single station with multiple servers to model all thread pool threads would inaccurately represent parallel execution of all requests. Conversely, a single server with multiple queues would overload the server even at low input rates. Empirical testing, reported in Sect. 4, demonstrated that the PAMSALPS proposed model, with a station per endpoint, yielded response times more closely aligned with real-world behavior across different workload rates. This configuration also mirrored actual system overload characteristics and simplified the modeling of complex microservice dependencies and interconnections.

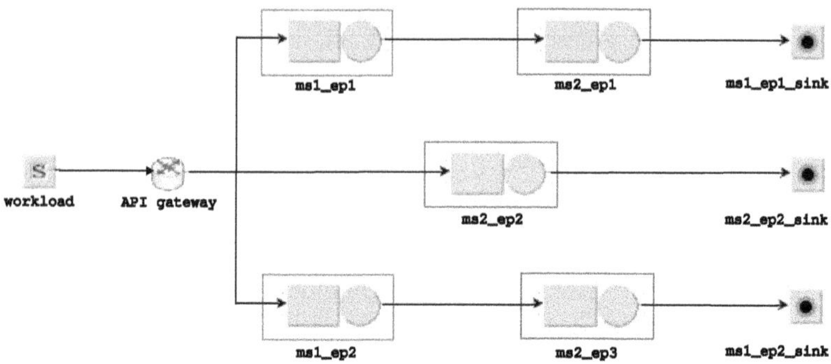


**Fig. 3.** Example of a QN modelling two microservice with five endpoints

Figure 3 shows a simple example of the PAMSALPS modeling technique on an API gateway MSA involving two interacting microservices ($m = 2$) ms1 (red boxed) and ms2 (blue boxed). The ms1 microservice exposes 2 endpoints ($e_1 = 2$), ms1_ep1 and ms1_ep2, whereas $ms2$ exposes 3 endpoints ($e_2 = 3$), ms2_ep1, ms2_ep2 and ms2_ep3. The API gateway MSA exposes 3 endpoints ($e = 3$. Thereby, the source node workload generates the workload and the API gateway node dispatches incoming requests to the corresponding externally exposed microservice endpoints, represented by the 3 sinks and related paths, respectively. Connections among the stations are directly and easily derived from the software project specifications; as in the example depicted in Fig. 3, microservices can be used for internal computation in complex systems, but it is easy to catch interdependencies. According to the proposed PAMSALPS modeling above, the response time of a request to the API gateway MSA includes all the delays from the API gateway router to the sinks, across the endpoint stations corresponding to the specific API endpoint invoked. For example, the response time for processing a request to the third API gateway endpoint is the sum of the time spent to visit both the stations ms1_ep2 and ms2_ep3. The PAMSALPS modeling approach allows to model the connections and data flows between microservices, including orchestrated workflows where a request to the API gateway endpoint internally triggers other requests to different endpoints.

The *model validation* is performed by comparing the performance metrics assessed by the model (e.g., throughput, latency), against the measured values from the benchmarking phase under similar workload conditions. The feedback from the validation, e.g. error estimates and similar metrics, can be used as feedback to the modeling step to improve the original model, as highlighted in Fig. 2 by the model validation-modeling loop. Once a final model is obtained, it is used to evaluate the actual workload condition and provide the appropriate parameters for the MSA endpoint capacity planning in the *model evaluation* stage. This model is thus exploited to assess scenarios by increasing system-wide workload. By observing the model behavior under these workloads, the optimal scalability thresholds and the required number of replicas for each microservice, grouping its endpoint replica demand, are determined. This process incorporates the cost implications of resource allocation, allowing to identify the configurations that balance performance requirements with resource efficiency. The output of this phase is a set of recommended scaling parameters for each microservice, to achieve optimal performance while minimizing resource costs. In the case of modifications in the modeled configuration, e.g. due to the release of a new version of a microservice with different performance, a feedback loop to the modeling step is activated to revise and update the original model accordingly.

Despite the proposed modeling approach effectively captures the behavior of single microservice endpoints, a scalability concern may arise since a microservice could expose tens of endpoints leading to combinatorial model complexity. This exponential growth in components could make simulation and analysis computationally unfeasible, challenging the practical application of this approach. However, a key observation mitigates this risk. In real-world systems, the primary scalability bottleneck often occurs with microservice endpoints that are widely shared across multiple service lines, i.e. MSA exposed endpoints. The number of such critical, high-traffic endpoints is usually limited, and orchestrated workflows are mainly sequential. Therefore, while a microservice may have many endpoints in total, only a manageable subset requires the granular, per-endpoint modeling we propose. This focus on the most impactful endpoints helps keep the model tractable and ensures the solution remains scalable for most practical microservice architectures.

### 3.3   Deployment

The final phase involves the implementation and enforcement of the scaling configurations obtained by the investigation phase within the actual MSA deployment environment. The number of replicas for each microservice endpoint is provisioned by the underlying orchestrator scaling capabilities (e.g. Kubernetes [10]), thus enforcing the *orchestration* step including the scaling management. The MSA behavior under real-world workloads, mirroring the high-workload scenarios investigated in the previous phase, is then carefully monitored by the *monitoring* step at endpoint level detail. The same key performance indicators of those collected during the initial benchmarking, are tracked to assess the effectiveness of the applied scaling configurations. In the case of changes affecting the endpoint, the monitored KPI, fed back to the model evaluation step, are thus evaluated by the latter. Thereby, a new configuration, adapting the microservice replicas (up- or down-scaling) to the current workload condition based on the endpoint demands is obtained and thus enforced by the orchestrator. This step also allows us to confirm whether the predicted performance improvements and resource utilization efficiencies materialize in the operational environment. Any discrepancies between the model predictions and the observed system behavior are analyzed to refine the model and the overall methodology. Successful validation confirms the efficacy of the proposed approach in achieving a performance-aware and cost-optimized scaling of the API gateway microservices architecture.

## 4   Case Study

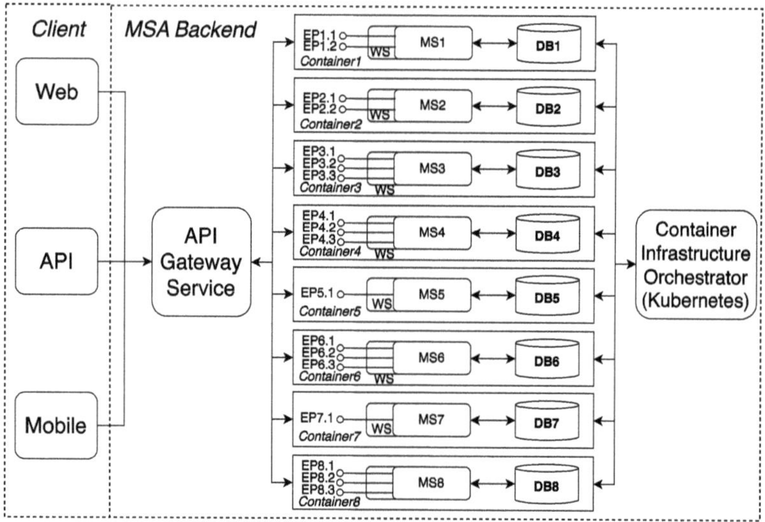


**Fig. 4.** Case study API GW MSA

This section describes the PAMSALPS approach, mainly focusing on the *investigation* phase, through an actual case study on automatic billing and invoicing process. It has been developed as a microservices architecture adopting technologies such as Java Springboot, Kubernetes and Docker for the deployment, and the Infrastructure as Code Terraform platform to manage the infrastructure hosted on the Microsoft Azure Cloud platform. A draft schema of the case study is provided in Fig. 4. The application implements a Gateway API pattern composed of eight microservices, each providing different endpoints. Overall, eighteen different endpoints are exposed by the six microservices; seven of them are used internally, implementing multi-step workflows. Thus, in the case study, only eleven endpoints are exposed by the API Gateway MSA Backend to the Client. This results from the above inception phase, which proceeds with a benchmarking of the case study MSA. Three workload conditions have been considered to such a purpose, as reported in Table 2 with the results thus obtained (*Real* columns).

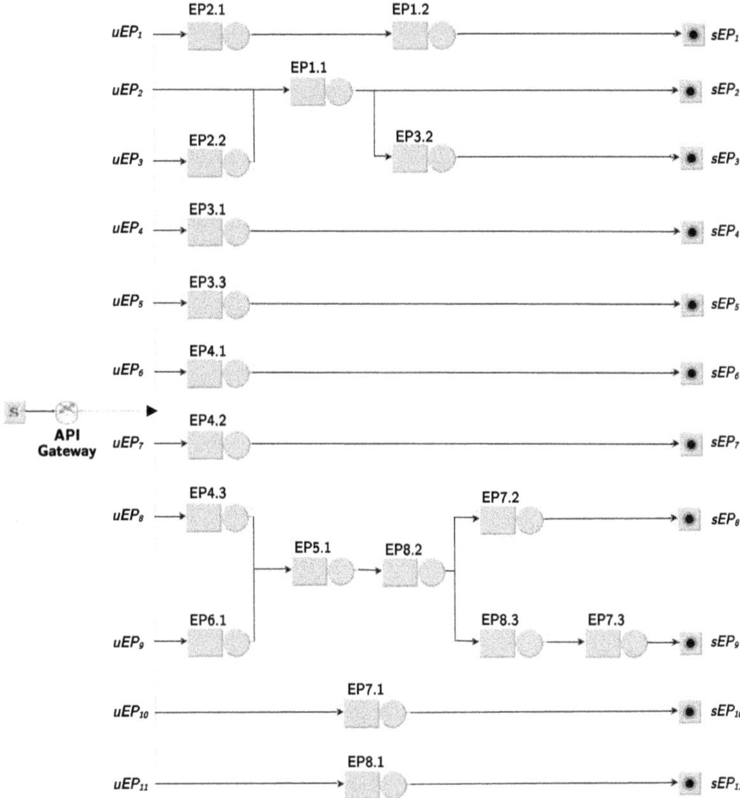


**Fig. 5.** Queuing Network model of the case study API Gateway MSA

Once the first inception phase of workload testing and data collection for each microservice is over, a queuing network model has been developed. By this model, the number of replicas for each microservice, based on different workloads, can be estimated. We experimented and tested the proposed approach by modeling the real system described in Fig. 4. The QN modeling the microservice architecture depicted in Fig. 4 is shown in Fig. 5. It has been developed from the workflow design of the microservice architecture and exploiting the modeling approach introduced in Sect. 3.2. It is worth noting that the QN model structure follows the software architecture of the real application. When a microservice calls another microservice internally, the output of the station modeling the microservice caller is connected to the queue modeling the called microservice. Similarly, the stations modeling microservices exposing end-points are all directly connected to the QN router which represents the API Gateway, while sinks are used to model the application outcomes. Requests to different endpoints are modeled with different classes generated by the API Gateway, and the appropriate station is selected by the API Gateway accordingly. As a consequence, the QN model overlies the logic of the software application. We denoted with $uEP_i$ and $sEP_i$, with $1 \le i \le 11$, the starting point of the user request submitted to the $i$-th end-point and its exit point, respectively. In this way, the sink names should correspond to the externally exposed endpoints. The model was created and analyzed using the Java Modeling Tool (JMT) [4]; it allowed us to obtain some indices, such as estimated response times, average utilization of each station, throughput, and other metrics.

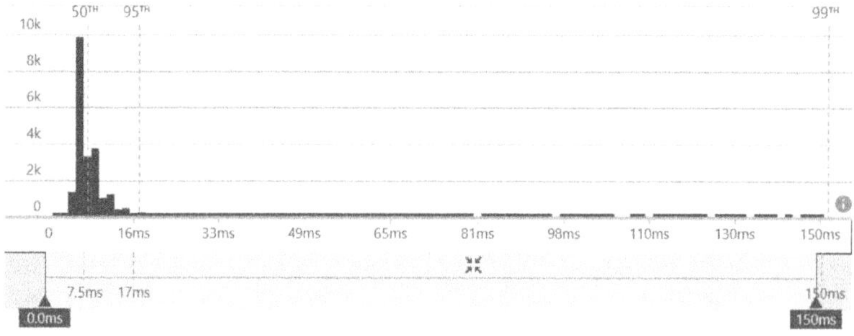


**Fig. 6.** Frequency histogram of the endpoint 1.2 of microservice MS1 (EP1.2) response time

In the benchmark phase, microservice endpoint response times have been measured and collected. This kind of metric can be obtained by the real system observing and measuring its behavior from an external access points. The case study on which we conducted the experimentation, based on the architecture shown in Fig. 4, consists of eight microservices $m = 8$, eighteen endpoints overall, and eleven endpoints ($e = 11$) exposed by the MSA API gateway service. As an example, Fig. 6 shows the frequency histogram of the microservice 1 endpoint 2 (EP1.2) response time. Despite the observed trend is not exponentially distributed, we chose to approximate EP1.2 and all other endpoint response times with exponentially distributed random variables, whose expected values have been obtained from experimental data, to exploit the Markovian theory in the QN assessment. This allows to exploit a simple analysis framework providing an acceptable approximation as demonstrated by the following experiments.

Thereby, the service times of the queuing network stations have been set to negative exponential distributions with expected values equal to the corresponding endpoint service time benchmarks. Specifically, the EP1.2 mean service time has been set to 8.03 $ms$, with rate $\lambda_{1.2} = 12.45$ $job/sec$. All the MSA case study endpoint service rate derived in the benchmark phase are reported in Table 1.

**Table 1.** Measured endpoints service rates

| Endpoint | EP1.1 | EP1.2 | EP2.1 | EP2.2 | EP3.1 | EP3.2 | EP3.3 | EP4.1 | EP4.2 |
|---|---|---|---|---|---|---|---|---|---|
| Service rate ($job/sec$) | 421.94 | 12.45 | 51.36 | 35.46 | 444.44 | 112.99 | 132.28 | 219.30 | 39.22 |
| Endpoint | EP4.3 | EP5.1 | EP6.1 | EP7.1 | EP7.2 | EP7.3 | EP8.1 | EP8.2 | EP8.3 |
| Service rate ($job/sec$) | 14.98 | 144.72 | 16.06 | 135.14 | 70.42 | 49.50 | 364.96 | 467.29 | 203.67 |

We then performed on the benchmark measurement of the average response times of the 11 API endpoints exposed by the MSA and of the average system time. The results of these experiments are shown in Table 2, compared against the values obtained by the QN model considering the single microservice endpoint service rates shown in Table 1. We assumed three different workloads setting the arrival rate of requests to $\lambda = 0.1,\ 1.0,\ 3.0$ $req/sec$. Each type of request has been supposed to be forwarded uniformly towards each exposed endpoint of a microservice. The columns labeled "*Real*" report the measured values, whereas the columns "*Model*" report the values estimated by the QN model.

**Table 2.** Measured vs model derived response times

| Requests arrival rate $\lambda$ (req/sec) | $\lambda = 0.1$ | | $\lambda = 1.0$ | | $\lambda = 3.0$ | |
|---|---|---|---|---|---|---|
| ***Response Time (ms)*** | *Real* | *Model* | *Real* | *Model* | *Real* | *Model* |
| *System* | 19.28 | 20.4 | 19.35 | 21.2 | 20.875 | 22.2 |
| $uEP_1$ | 2.37 | 2.35 | 2.81 | 2.38 | 2.57 | 2.4 |
| $uEP_2$ | 19.47 | 19.5 | 19.8 | 19.7 | 19.1 | 20.2 |
| $uEP_3$ | 28.2 | 28.2 | 28.6 | 28.4 | 27 | 29.3 |
| $uEP_4$ | 2.25 | 2.26 | 2.52 | 2.27 | 2.41 | 2.28 |
| $uEP_5$ | 7.56 | 7.58 | 7.77 | 7.61 | 6.88 | 7.65 |
| $uEP_6$ | 4.56 | 4.57 | 3.05 | 4.60 | 3.27 | 4.64 |
| $uEP_7$ | 25.47 | 25.4 | 26.5 | 26.3 | 34.4 | 27.6 |
| $uEP_8$ | 66.74 | 66.4 | 64.7 | 69.2 | 67.3 | 74.9 |
| $uEP_9$ | 62.25 | 62.5 | 63.2 | 63.8 | 70.6 | 66.3 |
| $uEP_{10}$ | 7.4 | 7.35 | 7.96 | 7.5 | 9.89 | 7.53 |
| $uEP_{11}$ | 2.74 | 2.76 | 2.61 | 2.74 | 2.32 | 2.75 |

It is possible to see that the data obtained in many endpoints are very similar, albeit with some margin of error. It is also worth noting that during the performed tests, a gradually increasing deterioration in performance was observed on some microservices, in particular on the microservices `microservice_5` and `microservice_7`. The latter behavior likely depends on the database schema to which the two microservices interface. Its fix is out of the scope of this work, and the proposed model does not take into account microservice degradation. All other endpoints can be considered reliable; it can be seen that the percentage error between the simulated and actual average response times ranges between 1% and 15%.

## 5   Conclusions

This paper introduces a novel approach to modeling microservice-based architectures by treating each microservice endpoint as a separate service station within a queuing network. This departs from traditional methods that model entire microservices as single entities. The core assumption is that each endpoint is served by at least one dedicated thread, allowing us to accurately capture resource utilization and performance bottlenecks at a finer granularity.

The resulting queuing network-based MSA capacity planning methodology demonstrates its potential to ensure scalability, efficiency, and cost-effectiveness across different workload conditions. The 3-step methodology, which estimates optimal microservice replica counts, has been indeed validated through empirical experimentation. The results indicate that the proposed modeling and evaluation approach outperforms direct empirical testing on real systems. While the validation phase involved a limited number of microservices, the methodology holds promise for broader application in large-scale, complex infrastructures.

Future research directions include extending the queuing network model to accommodate heterogeneous workloads and multi-datacenter deployments, as well as examining the influence of diverse deployment strategies on infrastructure performance and cost. Despite certain limitations, this work offers valuable insights for both practitioners and researchers concerned with scaling distributed microservice infrastructures within the domain of distributed systems and Cloud computing.

**Acknowledgements.** This work is partially funded by European Union - Next generation EU - PNRR - Missione 4, Componente 2, Investimento 1.1 - Bando PRIN 2022 PNRR - Decreto Direttoriale n. 1409 del 14-09-2022 - Progetto RESILIENT, CUP J53D23015040001, project id. P2022S4TTP, and Next generation EU - PNRR, SER-ICS – "SECURITY AND RIGHTS IN THE CYBERSPACE" project 3D-SEECSDE, CUP J33C22002810001, project id. PE00000014.

# References


1. Apache: Kafka. https://kafka.apache.org/
2. Bai, H., Xu, M., Ye, K., et al.: DRPC: distributed reinforcement learning approach for scalable resource provisioning in container-based clusters. IEEE Trans. Serv. Comput. **17**(2), 1071–1084 (2024). https://doi.org/10.1109/TSC.2023.3238314, https://www.scopus.com/record/display.uri?eid=2-s2.0-85199577365&origin=scopusAI
3. Balla, D., Simon, C., Maliosz, M.: Adaptive scaling of kubernetes pods. In: NOMS 2020-2020 IEEE/IFIP Network Operations and Management Symposium, pp. 1–5. IEEE (2020)
4. Bertoli, M., Casale, G., Serazzi, G.: JMT: performance engineering tools for system modeling. ACM SIGMETRICS Perform. Eval. Rev. **36**(4), 10–15 (2009)
5. Cao, J., Andersson, M., Nyberg, C., Kihl, M.: Web server performance modeling using an m/g/1/k*ps queue, vol. 2, pp. 1501–1506 (2003). https://doi.org/10.1109/ICTEL.2003.1191656
6. Cheng, K., Zhang, S., Tu, C., et al.: Proscale: proactive autoscaling for microservice with time-varying workload at the edge. IEEE Trans. Parallel Distrib. Syst. (2023). https://doi.org/10.1109/TPDS.2023.3234316, https://www.scopus.com/record/display.uri?eid=2-s2.0-85147290436&origin=scopusAI
7. Dragoni, N., et al.: Microservices: yesterday, today, and tomorrow. Present and Ulterior Software Engineering, pp. 195–216 (2017)
8. Elleithy, K.M., Komaralingam, A.: Using a queuing model to analyze the performance of web servers. In: International Conference on Advances in Infrastructure for e-Business, e-Education, e-Science, and e-Medecine on the Internet, Rome, Italy, pp. 21–27 (2002)
9. Gias, A.U., Casale, G., Woodside, M.: Atom: model-driven autoscaling for microservices. In: Proceedings - International Conference on Distributed Computing Systems, vol. 2019-July, pp. 1248–1258 (2019). https://doi.org/10.1109/ICDCS.2019.00127, https://www.scopus.com/record/display.uri?eid=2-s2.0-85074821654&origin=scopusAI
10. Google: Kubernetes. https://kubernetes.io/
11. Hasselbring, W.: Microservices for scalability: keynote talk abstract. In: Proceedings of the 7th ACM/SPEC on International Conference on Performance Engineering, pp. 133–134 (2016)
12. Incerto, E., Pizziol, R., Tribastone, M.: $\mu$opt: an efficient optimal autoscaler for microservice applications. In: Proceedings - 2023 IEEE International Conference on Autonomic Computing and Self-Organizing Systems, ACSOS 2023 (2023). https://doi.org/10.1109/ACSOS56966.2023.00028, https://www.scopus.com/record/display.uri?eid=2-s2.0-85181766241&origin=scopusAI
13. Khaleq, A.A., Ra, I.: Intelligent autoscaling of microservices in the cloud for real-time applications. IEEE Access **9**, 35464–35476 (2021)
14. Lakshya, Khan, M.I.H., Bajarange, A.P.: Enhancing cloud resource management through predictive neural networks and reinforcement learning. In: 2024 15th International Conference on Computing Communication and Networking Technologies, ICCCNT 2024 (2024). https://doi.org/10.1109/ICCCNT60019.2024.10440814, https://www.scopus.com/record/display.uri?eid=2-s2.0-85211189442&origin=scopusAI
15. Li, J., Li, S., Tan, J., et al.: DCScaler: spatiotemporal prediction aided distributed collaborative autoscaling of microservices. In: Proceedings - 2024 IEEE 10th International Conference on Edge Computing and Scalable Cloud, EdgeCom 2024

(2024). https://doi.org/10.1109/EdgeCom59651.2024.00027, https://www.scopus.com/record/display.uri?eid=2-s2.0-85202444940&origin=scopusAI

16. Liu, J., Zhang, S., Wang, Q.: $\mu$conadapter: reinforcement learning-based fast concurrency adaptation for microservices in cloud. In: SoCC 2023 - Proceedings of the 2023 ACM Symposium on Cloud Computing, pp. 83–95 (2023). https://doi.org/10.1145/3600756.3600765, https://www.scopus.com/record/display.uri?eid=2-s2.0-85178514610&origin=scopusAI

17. Lu, J., Gokhale, S.S.: Performance analysis of a web server. Int. J. Inf. Technol. Web. Eng. **3**(3), 50–65 (2008). https://doi.org/10.4018/JITWE.2008070104, https://doi.org/10.4018/jitwe.2008070104

18. Masse, M.: REST API Design Rulebook: Designing Consistent RESTful Web Service Interfaces. O'Reilly Media, Inc. (2011)

19. Nguyen, T.T., Yeom, Y.J., Kim, T., Park, D.H., Kim, S.: Horizontal pod autoscaling in kubernetes for elastic container orchestration. Sensors **20**(16), 4621 (2020)

20. Park, J., Jeong, B., Jeon, J., Jeong, Y.S.: Burst-aware horizontal autoscaling based on deep learning for stable microservices. Hum.-Cent. Comput. Inf. Sci. **14**(1), 1–22 (2024). https://doi.org/10.22937/HCIS.2024.14.1.1, https://www.scopus.com/record/display.uri?eid=2-s2.0-85210043275&origin=scopusAI

21. Pinciroli, R., Aleti, A., Trubiani, C.: Performance modeling and analysis of design patterns for microservice systems. In: 2023 IEEE 20th International Conference on Software Architecture (ICSA), pp. 35–46. IEEE (2023)

22. Rossi, F., Cardellini, V., Presti, F.L.: Hierarchical scaling of microservices in kubernetes. In: 2020 IEEE International Conference on Autonomic Computing and Self-Organizing Systems (ACSOS), pp. 28–37. IEEE (2020)

23. Slothouber, L.P., et al.: A model of web server performance. In: Proceedings of the 5th International World Wide Web Conference (1996)

24. Xie, S., Wang, J., Li, B., et al.: PBScaler: a bottleneck-aware autoscaling framework for microservice-based applications. IEEE Trans. Serv. Comput. **17**(1), 337–350 (2024). https://doi.org/10.1109/TSC.2023.3233618, https://www.scopus.com/record/display.uri?eid=2-s2.0-85188427054&origin=scopusAI

25. Yeager, N.J., McGrath, R.E.: Web Server Technology. Morgan Kaufmann (1996)

26. Zhao, J., Jing, S., Jiang, L.: Management of API gateway based on micro-service architecture. In: Journal of Physics: Conference Series, vol. 1087, p. 032032. IOP Publishing (2018)

27. Zheng, H., Kramer, J., Chang, S.K.: Software design pattern analysis for microservices architecture using queuing networks (2021)

# On Synchronous Approximations of Non-Markovian Processes with Dual Bounds

Reinhard German$^{(\boxtimes)}$ 

Department of Computer Science 7 (Communication Systems and Computer Networks), Friedrich-Alexander-Universität Erlangen-Nürnberg, Erlangen, Germany
reinhard.german@fau.de

**Abstract.** A concept for the approximate steady-state analysis of non-Markovian processes is suggested, in which deterministically timed activities are allowed to take place concurrently. The approach works by mapping the actual process on a simplified process in which the deterministic timing is synchronized and the process can be solved by using an embedded Markov chain. In the simplified process the deterministic activities can be made faster or slower than in the actual process such that upper and lower bounds for the resulting state probabilities can be given. This is possible without state-space expansion and leads already to good results in the reported experiments. However, if the gap between the bounds is not sufficient, a phase expansion can be used, in which the synchronous approximation error is restricted to just one phase. The approach is explored for an M/D/2/K queuing system but is not restricted to it. Traditional expansion by phase-type distributions does not provide bounds.

**Keywords:** Non-Markovian Processes · Approximate Analysis · Bounds

## 1 Introduction

Non-Markovian stochastic processes constitute a model class well-known from literature. In the continuous domain, besides exponentially timed memoryless activities also non-Markovian ones characterized by general distributions are allowed. This might be relevant for many modeling situations in which the memoryless assumption is inadequate, such as the Weibull distribution in reliability analysis. A special case is a combination of activities with exponentially and deterministically timed durations. In the domain of *stochastic Petri nets* (SPNs) [15] this has been known as *deterministic and stochastic Petri nets* [16], however, the analysis is independent of the SPN modeling formalism. Under the so-called *enabling condition* at most one non-Markovian activity is allowed in each state, meaning that they must not take place concurrently. For steady-state analysis, Markov renewal theory (MRT) [4] can be used, as has been suggested in [16]


L. Carnevali and J. Doncel (Eds.): EPEW 2025, LNCS 15657, pp. 115–130, 2026.
https://doi.org/10.1007/978-3-032-16345-5_9

and enhanced in many other works, e.g., [14]. It can easily be generalized to deal with general distributions instead of just deterministic times, e.g., [5,9]. The MRT approach requires the enabling condition to hold, with few exceptions in special cases.

Alternatively, *supplementary age variables* can be used, as described originally in [7] and enhanced in, e.g., [9]. A third approach is very well known: approximating general distributions by *phase-type distributions*, constituted by a continuous-time Markov chain with an absorbing state, as originally described in [8] and enhanced in many works, e.g., in [12]. In case of deterministic timing it is known that the most efficient approximation inside the class of phase-type distributions is the *Erlang distribution* [1], the accuracy of the approximation is however not known in advance. Discrete phase type expansions have also been considered, see, e.g., [6,12]. By generalizing phase-type distributions to *matrix exponential distributions* [12], where the interpretation of phases is relaxed, more degrees of freedom exist, especially *concentrated matrix exponential distributions* (CMEs) [13] allow for lower variances than the Erlang distribution with the same order, and is still applicable for analysis. The parameters must be obtained by numerical optimization-based procedures. Alternative approaches use either *generalized semi-Markov processes* [14] (has demonstrated to work with the $M/D/2/K$ system) or *stochastic state classes* [18] (has limits to deal with cyclic structures). All approaches suffer when the enabling condition is not met, since then the memories of the concurrent non-Markovian activities have to be represented, either by supplementary variables or by phase expansions. Two other approaches can be mentioned: *cascaded DSPNs* described in [9] when deterministic times are multiples of each other and certain structural restrictions hold, the analysis can be traced back to MRT and *hierarchical semi-Markov processes with parallel regions* (HSMPs) [10] where hierarchical conjunctions and disjunctions can be leveraged for an efficient analysis, it has been extended in [2] by the time advance mechanism known from stochastic state classes to take exits in parallel regions with different time origins correctly into account and in [3] it was put on a formal semantical basis, extended further, and implemented in the Pyramis library.

Here a concept is explored in which non-Markovian processes can be approximated by mapping the actual process on a simplified process which is subject to MRT analysis. In the mapping the concurrent activities are synchronized in such a way, that the simplified process is either faster or slower. This allows to compute bounds for the state probabilities and hence also for the measures derived by them. In case of two identically deterministically timed activities no state expansion is needed for that. This might already be sufficient in many modeling situations. Alternatively, synchronous phase expansions can be used in order to improve the bounds in an encapsulation approach: improve as long as the gap between the bounds is too large for the modeling question. In case of non-identically timed activities also a phase expansion would be needed. We demonstrate the concept for the $M/D/2/K$ system, it is however not restricted to it and a more general theory needs to be developed based on it. Whether the approximations based on accelerating and slowing down activity durations as proposed are really leading to bounds in more general cases is not yet clear and must be validated in further work, it also depends on stochastic monotonicity [17].

We show the synchronous approximation without state expansion in Sect. 2, the synchronous approximation based on phase expansions in Sect. 3, numerical results in Sect. 4, and conclusions in Sect. 5.

## 2   Approximations Without State Expansion

In this chapter we describe the approach of synchronous approximations by the example of the $M/D/2/K$ queuing system. We start with a description of the stochastic process shown in Fig. 1: Poisson arrivals with rate $\lambda$ (in case of an arrival to an empty system one of the servers will be chosen with equal probability of $1/2$), two identical servers with a deterministic service time $\tau$, and a capacity of $K$ customers including the servers. In the state space arabic numbers give the number of customers in the system, capital letters whether service unit $A$ or $B$ is busy in case of one customer in the system. Arrivals are labeled with the rate. $x$ stands for the time server $A$ has been busy, and $y$ is the same for server $B$ (i.e., $x$ and $y$ are *supplementary age variables*). The age variables are initiated with zero in the instant of service start and increase with the same constant speed. When one of them reaches the value of $\tau$, a service is completed, a state transition to the left happens and the age variable is reset to zero if the service restarts (if the destination state still has at least two customers), in case of state transitions caused by arrivals the age variables keep their values. The labels of the state transitions caused by service completions are Dirac impulses $\delta(x - \tau)$ for unit $A$ and $\delta(y - \tau)$ for unit $B$. The Dirac impulse can be considered as a generalized hazard rate in the deterministic case. The complexity of this process mainly comes from the fact that the second service can start at any value of the age variable of the other service, therefore all possible combinations of the two real variables are possible and must be taken into account in the analysis.

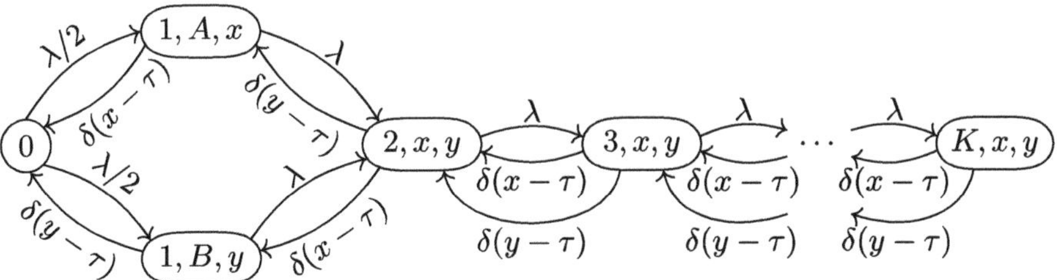


**Fig. 1.** Stochastic process underlying an $M/D/2/K$ system.

## 2.1   A Faster Approximation

Now we show a possibility to synchronize both services such that they always start at the same time, leading to a stochastic process in which at most one non-Markovian activity is possible at any time and the analysis gets simpler. This of course changes the actual process. However, we change it in such a way that the services happen faster than in the actual system. In order to get a close approximation we try to make it just as little faster as possible. In order to keep the solution efficient, we also do not want to add additional states.

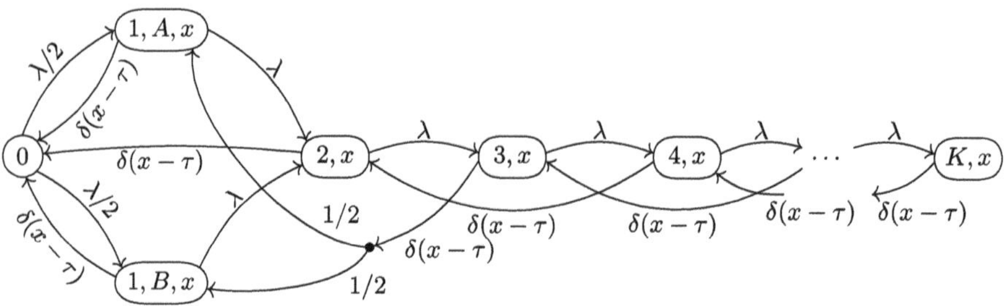


**Fig. 2.** Synchronous faster approximation of the $M/D/2/K$ system.

Figure 2 shows the process which is a synchronized faster approximation of an $M/D/2/K$ system. The states are the same as in the actual process besides the fact that we have age variable $x$ instead of $y$ in state $(1, B, x)$ and that we just have one age variable $x$ in all states with at least 2 customers in the system. Furthermore, when the age variable $x$ reaches the value of $\tau$ with both servers busy, a state transition happens to a state with two customers less, corresponding to two service completions taking place in the same instant of time: both services complete in a synchronized manner when $x$ reaches $\tau$ (labeled by $\delta(x - \tau)$); in case of 2 customers in the system the synchronized services lead back to an idle system, in case of 3 customers in the system the synchronized services lead to a system with one busy server (since the information of which one started later is lost, a probabilistic choice is needed to decide whether to go either to state $(1, A, x)$ or to $(1, B, x)$, without consuming time, for this, a *pseudo state*, as known from statecharts, is used and visualized as a black dot), and in case of $i \geq 4$ customers in the system, the synchronized services lead to a system with $i - 2$ customers. Since both service units work in the same way, it would be possible to drop the information of which server is busy and to represent them in just one state, it is however not done here in order to better illustrate the approximation.

The faster approximation can be understood by the following argument. Let's assume that at time $t$ a customer arrives to an idle system and that after time $0 < a < \tau$ another customer arrives. The first customer will be served at $t+\tau$, as in the actual process. However, the second customer will be served at the same time just after $\tau - a$ time units, which is faster than in the actual process. All other customers arriving during the same busy period with both servers busy will receive the correct service time equal to $\tau$; if for some time just one server is busy, the above argument is repeated and as a consequence, all service times are either correct or faster.

This process can be analyzed by using MRT in a standard way, we use the notation of [9]. Let $\mathcal{S} = \{0, 1, ..., K + 1\}$ be the state space (note that this numbering is slightly different than in the figure), where 0 stands for the idle system, 1 for one customer in the system at server $A$, 2 for one customer in the system at server $B$, and $3 \leq i \leq K + 1$ for $i - 1$ customers in the system. Let further $\mathcal{S}^E = \{0\}$ be the set in which only exponential activities are possible and let $\mathcal{S}^d = \{1, 2, ..., K + 1\}$ be the set in which a deterministic service is possible. Regenerations of the stochastic process are defined as the next customer arrival to an idle system ($i = 0$) and as the next service for all states $i \in \mathcal{S}^d$.

Let all exponential state transitions be gathered in matrix $\mathbf{Q}$, which is formally a generator matrix of a *continuous time Markov chain* (CTMC). Let all state transitions after a service completion (the *switching probabilities*) be represented in matrix $\boldsymbol{\Delta}$ (e.g., $\delta_{3,0} = 1$ and $\delta_{4,1} = \delta_{4,2} = 1/2$). Let furthermore for each subset $\mathcal{S}^A \subseteq \mathcal{S}$ a *filter matrix* $\mathbf{I}^A$ be defined as:

$$I_{ij}^A = \begin{cases} 1 & i = j, i \in \mathcal{S}^A, \\ 0 & \text{otherwise.} \end{cases}$$

With the filter matrix we can write $\mathbf{Q}^E = \mathbf{I}^E \mathbf{Q}$ and $\mathbf{Q}^d = \mathbf{I}^d \mathbf{Q}$ containing just rows corresponding to the set indicated by the superscript. We further need transient analysis of state changes according to arrivals when a service is ongoing, matrix $\boldsymbol{\Omega}$ contains all transient probabilities in the instant before a service happens and matrix $\boldsymbol{\Psi}$ contains the conditional mean sojourn times until the service:

$$\boldsymbol{\Omega} = \mathbf{I}^d e^{\mathbf{Q}^d \tau}, \quad \boldsymbol{\Psi} = \mathbf{I}^d \int_0^\tau e^{\mathbf{Q}^d x} \, dx.$$

The matrix exponential and its integral can be computed iteratively by *uniformization*, for which *Poisson probabilities* are needed and the numerical error can be bounded. The stochastic matrix $\mathbf{P}$ of the *embedded Markov chain* (EMC) at regeneration instants and the matrix $\mathbf{C}$ of conditional sojourn times in states between two regenerations are given by:

$$\mathbf{P} = \mathbf{I}^E - \text{diag}^{-1}\left(\mathbf{Q}^E\right)\mathbf{Q}^E + \boldsymbol{\Omega}\boldsymbol{\Delta}, \quad \mathbf{C} = -\text{diag}^{-1}\left(\mathbf{Q}^E\right) + \boldsymbol{\Psi},$$

where $\text{diag}^{-1}\left(\mathbf{Q}\right)$ is the matrix which contains the inverse elements of the diagonal of $\mathbf{Q}$ and zeroes at all other entries. The stationary solution $\mathbf{u}$ of the EMC is obtained by solving the linear system $\mathbf{u} = \mathbf{u}\mathbf{P}, \mathbf{u}\mathbf{e} = 1$ and the solution $\boldsymbol{\pi}$ of the faster approximation of the $M/D/2/K$ is finally obtained by normalization with the conditional sojourn times: $\boldsymbol{\pi} = \frac{\mathbf{u}\mathbf{C}}{\mathbf{u}\mathbf{C}\mathbf{e}}$ ($\mathbf{e}$ is a vector of ones).

## 2.2   A Slower Approximation

Now we are looking for a possibility to synchronize both services such that the services happen more slowly than in the actual process. Again we want to keep the process as simple as possible and to have it as little slower as possible.

**Case 1 (Additional States):** One possibility is to add further states before the actual service starts. Figure 3 shows such a synchronous slower approximation which needs additional states, leading almost to a doubling of the state space size. For every state with at least two customers in the system, two states are needed: $(i, -1, x)$ means that $2 \le i \le K$ customers are in the system and the service has not yet started and in $(i, 0, x)$ the service is ongoing.

The argument why this is slower is as follows. Assume that at time $t$ a first service has started and that after time $0 < a < \tau$ another customer arrives. The result is a transition to state $(2, -1, x)$ in which the already started service is finished without effect. After $x$ has reached $\tau$ a transition to state $(2, 0, x)$ takes place, the first service is repeated and the second one is started new. Thus, the first service needs $2\tau$ and the second service needs $2\tau - a$. If in state $(2, -1, x)$ further customers arrive, state changes to $(i, -1, x), 2 < i \le K$, happen and when $x$ has reached $\tau$ a transition to state $(i, 0, x)$ takes place. All further arriving customers in a busy period with both servers busy will therefore get the service time $\tau$ as in the actual process. If for some time just one server is busy, the above argument is repeated and as a consequence, all service times are either correct or slower. For a solution the procedure explained for the faster approximation can be used, with adapted sets $\mathcal{S}$ and $\mathcal{S}^d$ as well as matrices $\mathbf{Q}$ and $\mathbf{\Delta}$.

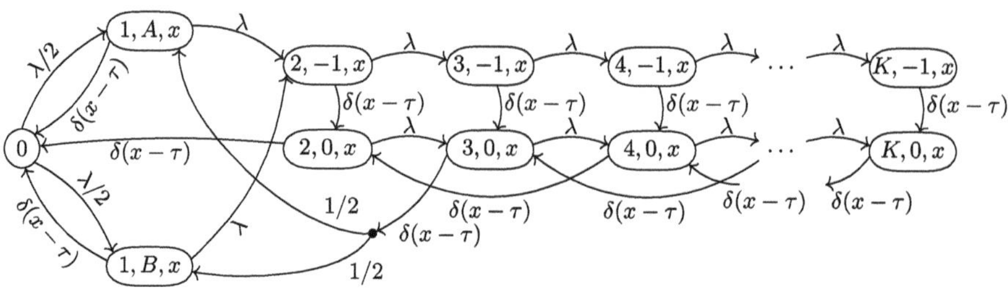


**Fig. 3.** Synchronous slower approximation of the $M/D/2/K$ system using additional states.

**Case 2 (Preemptions):** A synchronous slower approximation without additional states is also possible by taking advantage of *preemptions*. Preemptions are customer arrivals which reset the age variable. This is shown in Fig. 4. The rates from states $(1, A, x)$ and $(1, B, x)$ to $(2, x)$ shown in Fig. 2 are now assumed to be preemptions and drawn as thick lines in Fig. 4: the service restarts with an age variable set to zero after these transitions.

The slower approximation can be understood by the following argument. Let's assume that at time $t$ a customer arrives to an idle system and that after time $0 < a < \tau$ another customer arrives. The first service restarts and the second service starts at time $t + a$, therefore both services will be completed at time $t + a + \tau$. Hence the first service needs $a + \tau$ time units, which is slower than in the actual process, the second needs $\tau$ time units, as in the actual process. All other customers arriving during the same busy period with both servers busy will receive the correct service time $\tau$. If for some time just one server is busy, the above argument is repeated and as a consequence, all service times are either correct or slower.

For the analysis preemptions are also taken as regenerations and can be represented in matrix $\bar{\mathbf{Q}}$, here $\bar{q}_{1,3} = \bar{q}_{2,3} = \lambda$, all other entries are equal to zero; the preemptions must be taken out of $\mathbf{Q}$, but $\mathbf{Q} + \bar{\mathbf{Q}}$ must constitute a valid generator matrix of a CTMC. The stochastic matrix $\mathbf{P}$ has then an additional term corresponding to the preemptions: $\mathbf{I}^E - \mathrm{diag}^{-1}\left(\mathbf{Q}^E\right)\mathbf{Q}^E + \mathbf{\Omega\Delta} + \mathbf{\Psi}\bar{\mathbf{Q}}$, all remaining steps for the solution are identical to the previous case. For an explanation of these computations including preemptions see, e.g., [9].

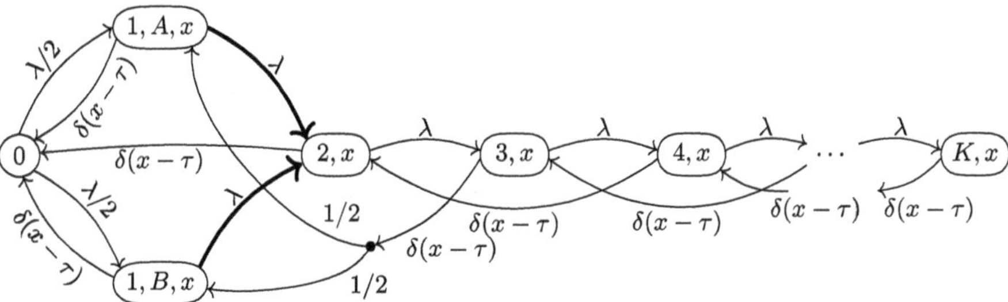


**Fig. 4.** Synchronous slower approximation of the $M/D/2/K$ system using preemptions.

The second approximation based on preemptions is simpler than case 1: it does not need additional states and uses the same topology as the faster approximation, just some exponential state transitions must be taken as preemptions. It is also faster than case 1, as numerical experiments will show, it provides therefore a better approximation quality and should be preferred if no expansion is used. The first approximation based on additional states is however also valid and provides the basis for a slower approximation with phase expansions as shown in the next chapter.

## 3   Synchronous Phase Expansions

In order to improve the gap between the two approximations a synchronous phase expansion is possible, meaning that one of the two deterministic services is represented by a sum of smaller deterministic phase times and the principles of faster and slower are applied just to a phase and not to the overall service time. By increasing the number of phases the error is decreased but of course also the states space grows. The hope is that already with few phases a sufficient encapsulation for a bounded error can be found.

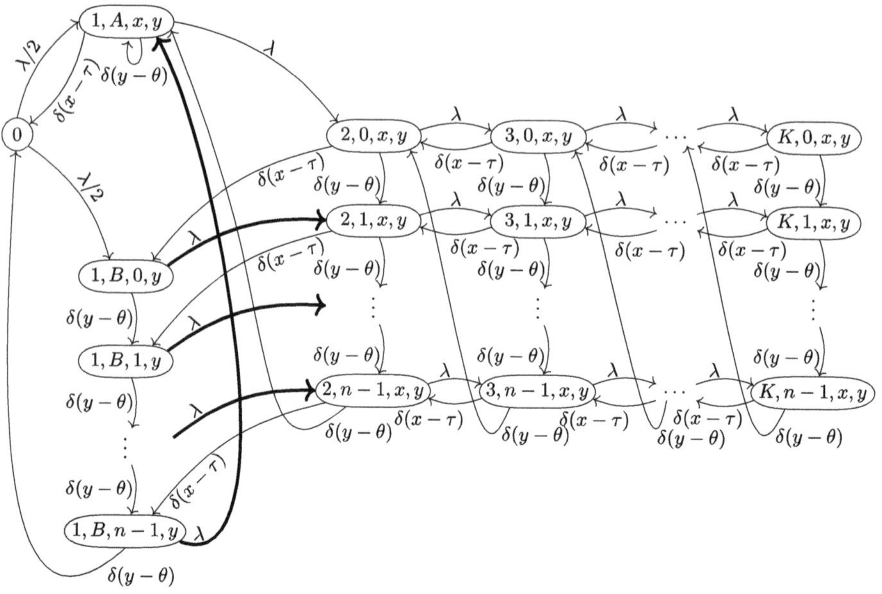

**Fig. 5.** Synchronous faster expanded approximation of the $M/D/2/K$ system.

## 3.1   Phase Expansion for a Faster Approximation

Figure 5 shows a faster expanded approximation of the $M/D/2/K$ system with a phase duration of $\theta$ with $\tau = n\theta$. The service time of unit $A$ is represented directly by its age variable $x$, whereas the service time of unit $B$ is subdivided into phases with an age variable $y$. The states $(1, B, j, y)$, $0 \le j \le n-1$ represent a system with one customer in the system in unit $B$ with $j$ phases completed. The states $(i, j, x, y)$, $2 \le i \le K, 0 \le j \le n - 1$, represent a system with at least two customers in the system with $j$ phases completed. When $y$ reaches $\theta$ and when $x$ reaches $\tau$ state transitions happen as shown in the figure by the arcs labeled with Dirac impulses. Note that in state $(1, A, x, y)$ a phase transition is possible (leading to a self-loop), although service unit $B$ is not busy. It is added artificially, when one or more further arrivals take place, the elapsed time in that phase is taken into account in states $(i, 0, x, y)$, $2 \le i \le K$ and make the first phase transition faster. If no arrival happens, the phase is finished without effect and restarts again. To have phase transitions and services synchronized, arrivals in states $(1, B, j, y)$, $0 \leq j \leq n - 1$, preempt the current phase and lead to the next phase immediately (to states $(2, j + 1, x, y)$, $0 \leq j < n - 1$), and to state $(1, A, x, y)$, respectively), making the system faster. When both a phase of unit $B$ and service of unit $A$ finish at the same time we assume that first the state change of the phase completion and only then that of the service takes place (immediately without time consumption). Because of this unit $A$ completions in the last row (in states $(i, n - 1, x, y)$, $2 \leq i, \leq K$) are actually not possible. Preemptions are again shown as thick lines.

For the solution the principle of *cascaded embedding* described in [9] can be adapted. We define sets $\mathcal{S}^E = \{0\}$ and

$$\mathcal{S}^A = \{\text{states supplemented with age variable } x\},$$
$$\mathcal{S}^B = \{\text{states supplemented with age variable } y\}, \mathcal{S}^{\hat{B}} = \mathcal{S}^B \setminus \mathcal{S}^A.$$

Note that in states of $\mathcal{S}^A$ both service completions of unit $A$ and phase transitions are possible and in states of $\mathcal{S}^{\hat{B}}$ just phase transitions are possible. The switching probabilities for deterministically timed transitions are collected in $\delta_{i,j}^X$, where $X = A$ stands for service completions of unit $A$ (i.e., when $x$ reaches $\tau$ in states of $\mathcal{S}^A$), $X = BA$ stands for phase transitions when unit $A$ is busy (i.e., when $y$ reaches $\theta$ in states of $\mathcal{S}^A$), and $X = \hat{B}$ stands for phase transitions when only unit $B$ is busy (i.e., when $y$ reaches $\theta$ in states of $\mathcal{S}^{\hat{B}}$). To deal with preemptions, diagonal entries of states outside $\mathcal{S}^X$ need to be set to one in all cases.

Transient state probabilities are collected in matrix $\boldsymbol{\Omega}^A(\theta)$ in the instant just before a phase change happens, $\tilde{\boldsymbol{\Omega}}^A(\theta)$ in the instant after the phase change, and $\tilde{\boldsymbol{\Omega}}^A$ after a service of unit $A$ has been completed:

$$\boldsymbol{\Omega}^A(\theta) = \mathbf{I}^A e^{\mathbf{Q}^A \theta}, \ \tilde{\boldsymbol{\Omega}}^A(\theta) = \boldsymbol{\Omega}^A(\theta)\boldsymbol{\Delta}^{BA}, \ \tilde{\boldsymbol{\Omega}}^A = \left(\tilde{\boldsymbol{\Omega}}^A(\theta)\right)^n \boldsymbol{\Delta}^A.$$

Conditional mean sojourn times are given in matrix $\boldsymbol{\Psi}^A(\theta)$ during a phase transition and $\boldsymbol{\Psi}^A$ during a service of unit $A$:

$$\boldsymbol{\Psi}^A(\theta) = \mathbf{I}^A \int_0^\theta e^{\mathbf{Q}^A t} \, dt \, \mathbf{I}^A, \ \boldsymbol{\Psi}^A = \sum_{j=1}^{n-1} \left(\tilde{\boldsymbol{\Omega}}^A(\theta)\right)^j \boldsymbol{\Psi}^A(\theta).$$

When just unit $B$ is enabled, probabilities of state transitions before a phase change and after a phase change are given by $\boldsymbol{\Omega}^{\hat{B}}$ and $\tilde{\boldsymbol{\Omega}}^{\hat{B}}$, and the conditional mean sojourn times by $\boldsymbol{\Psi}^{\hat{B}}$:

$$\boldsymbol{\Omega}^{\hat{B}} = \mathbf{I}^{\hat{B}} e^{\mathbf{Q}^{\hat{B}} \theta}, \ \tilde{\boldsymbol{\Omega}}^{\hat{B}} = \boldsymbol{\Omega}^{\hat{B}} \boldsymbol{\Delta}^{\hat{B}}, \ \boldsymbol{\Psi}^{\hat{B}} = \mathbf{I}^{\hat{B}} \int_0^\theta e^{\mathbf{Q}^{\hat{B}} t} \, dt \, \mathbf{I}^{\hat{B}}.$$

As described above we assume that when a phase transition of unit $B$ and a completion of service $A$ is possible at the same time, we assume that first the completion of the phase and only then that of the service takes place, as is represented in the matrices of switching probabilities; this assumption could be

124    R. German

dropped by using an extra matrix of switching probabilities for both completions when $x$ reaches $\tau$. The matrices can be added:

$$\tilde{\boldsymbol{\Omega}} = \tilde{\boldsymbol{\Omega}}^A + \boldsymbol{\Omega}^{\hat{B}}, \quad \boldsymbol{\Psi} = \boldsymbol{\Psi}^A + \boldsymbol{\Psi}^{\hat{B}}$$

and inserted in the stochastic matrix $\mathbf{P}$ of the EMC and in matrix $\mathbf{C}$ of conditional sojourn times (by taking preemptions into account), thus the previous solution procedure can be applied again. State probabilities of expanded states have to be summed up.

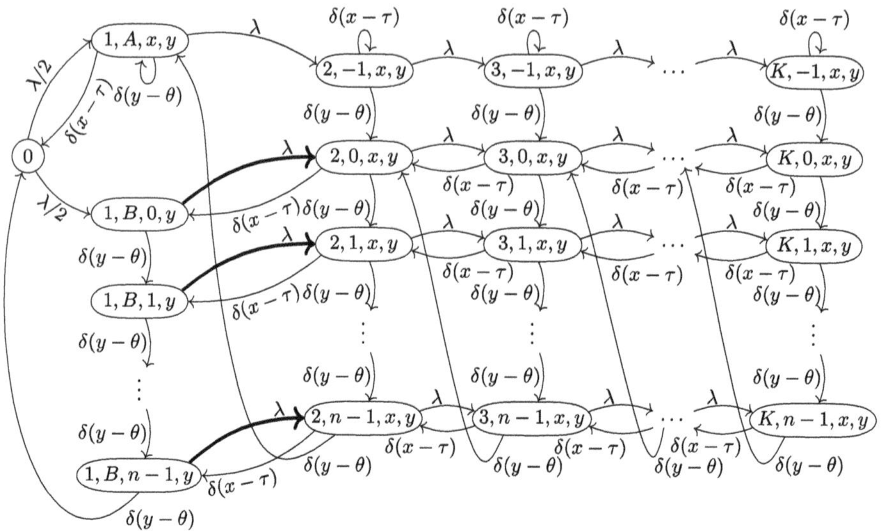

**Fig. 6.** Synchronous slower expanded approximation of the M/D/2/K system.

## 3.2    Phase Expansion for a Slower Approximation

Figure 6 shows a slower expanded approximation of the $M/D/2/K$ system. The states $(i, j, x, y)$, $2 \leq i \leq K, -1 \leq j \leq n-1$, represent a system with at least two customers in the system with $j$ phases completed, $j = -1$ is an additional phase before the service can start, to make it slower, following the concept as explained for the slower approximation without phase expansion but with additional states in Sec. 2.2 (case 1). For slower phase transitions after arrivals in states $(1, B, j, y)$, $0 \leq j \leq n-1$, these preempt the current phase and get to the same phase immediately (to states $(2, j, x, y)$, $0 \leq j < n-1$) and restart it. Note that in states $(i, -1, x, y)$, $2 \leq i \leq K$ self-loops with $\delta(x - \tau)$ are needed for formal reasons for the correct definitions of the matrices, however, these transitions can never happen because a phase completion of unit $B$ happens before a completion of unit $A$ and therefore these states will be left before the age variable $x$ reaches

$\tau$. Similarly, unit $A$ completions in states $(i, n-1, x, y)$, $2 \leq i \leq K$) are actually not possible. By adapting the reachability sets and all matrices as indicated in the figure, a solution is possible following the same procedure as in the faster expanded approximation.

## 4   Numerical Experiments

Numerical experiments were conducted using *SPNica*, a Mathematica library, described in [9], updated for the current version of Mathematica[1]. CME[2] computations [13] were done with the Mathematica version of BuTools[3] and a matrix-geometric formulation for the $M/D/2/K$ queuing system in which the deterministic services are approximated by CMEs. We examine the queuing system with low and medium utilization ($\lambda = 0.1$ and $\lambda = 0.5$, $\tau = 1, K = 10$) and give the probability of an idle system $P(idle)$, a full system $P(full)$, and the mean response time $R$ computed with Little's law. We compare these values for various ways of approximation in Table 1:

- The first four rows show results when the deterministic service is replaced by an exponential distribution (named "Exponential") and by Erlang distributions with 2, 10, and 100 phases (named "Erlang 2, 10, 100", respectively).
- The next three rows show results when concentrated matrix exponentials (CMEs) with orders of 5, 10, and 15 are used (named "CME 5, 10, 15", respectively). The order defines the enlargement of the state space such as the number of phases of a phase type distribution. The parameters of the CMEs have been chosen such that the squared coefficient of variation for CMEs with orders of 5, 10, 15 is $0.026, 0.0057, 0.0024$, which would correspond to Erlang distributions with $39, 176, 417$ phases, respectively.
- The next row shows the results for the faster approximation without phase expansion (named "faster w/o"). The next three rows show the results for the faster approximation with an expansion by 2, 10, and 100 phases (named "faster 2, 10, 100", respectively).
- The next four rows show the results for the slower approximation with an expansion by 2, 10, and 100 phases (named "slower 2, 10, 100", respectively). "slower 1" stands for the case of a slower approximation when additional states are used but no further expansion takes place, as described in Sect. 2.2 (first case). The last row shows the results for the slower approximation based on preemptions without phase expansion as described in Sect. 2.2 (second case), named "slower w/o".

Due to restricted capacities of Mathematica in case of 100 phases a Python script using sparse matrix structures was used and both servers were approximated by an Erlang distribution. With BuTools the results for the Erlang distribution with all phases were confirmed.

---

[1] https://github.com/cs7org/MD2K-SynchronousApproximation.
[2] https://github.com/ghorvath78/iltcme.
[3] https://webspn.hit.bme.hu/~telek/tools/butools/.

**Table 1.** Results for $\lambda = 0.1$ and $\lambda = 0.5$, $\tau = 1$, $K = 10$

| approach | $\lambda = 0.1$ | | | $\lambda = 0.5$ | | |
|---|---|---|---|---|---|---|
| | $P(idle)$ | $P(full)$ | $R$ | $P(idle)$ | $P(full)$ | $R$ |
| Exponential | 0.90476 | $1.7671 \cdot 10^{-13}$ | 1.0025 | 0.60000 | $1.1444 \cdot 10^{-6}$ | 1.0667 |
| Erlang 2 | 0.90475 | $8.0644 \cdot 10^{-15}$ | 1.0020 | 0.59921 | $1.1658 \cdot 10^{-7}$ | 1.0522 |
| Erlang 10 | 0.90472 | $1.3242 \cdot 10^{-16}$ | 1.0017 | 0.59757 | $5.4622 \cdot 10^{-9}$ | 1.0409 |
| Erlang 100 | 0.90470 | $3.5470 \cdot 10^{-17}$ | 1.0016 | 0.59670 | $2.0677 \cdot 10^{-9}$ | 1.0383 |
| CME 5 | 0.90470 | $4.5048 \cdot 10^{-17}$ | 1.0016 | 0.59688 | $2.5044 \cdot 10^{-9}$ | 1.0387 |
| CME 10 | 0.90470 | $3.3823 \cdot 10^{-17}$ | 1.0016 | 0.59663 | $1.9816 \cdot 10^{-9}$ | 1.0381 |
| CME 15 | 0.90470 | $3.2592 \cdot 10^{-17}$ | 1.0016 | 0.59658 | $1.9134 \cdot 10^{-9}$ | 1.0380 |
| faster w/o | 0.90870 | $2.9831 \cdot 10^{-17}$ | 0.9588 | 0.64374 | $1.5779 \cdot 10^{-9}$ | 0.9042 |
| faster 2 | 0.90580 | $1.6931 \cdot 10^{-17}$ | 0.9784 | 0.61251 | $1.1874 \cdot 10^{-9}$ | 0.9483 |
| faster 10 | 0.90491 | $2.5791 \cdot 10^{-17}$ | 0.9970 | 0.59960 | $1.6586 \cdot 10^{-9}$ | 1.0199 |
| faster 100 | 0.90472 | $3.1230 \cdot 10^{-17}$ | 1.0011 | 0.59685 | $1.8468 \cdot 10^{-9}$ | 1.0361 |
| slower 100 | 0.90468 | $3.2678 \cdot 10^{-17}$ | 1.0016 | 0.59624 | $1.8941 \cdot 10^{-9}$ | 1.0386 |
| slower 10 | 0.90448 | $4.0347 \cdot 10^{-17}$ | 1.0020 | 0.59358 | $2.1377 \cdot 10^{-9}$ | 1.0448 |
| slower 2 | 0.90364 | $1.0171 \cdot 10^{-16}$ | 1.0036 | 0.58279 | $4.0555 \cdot 10^{-9}$ | 1.0751 |
| slower 1 | 0.90009 | $2.2469 \cdot 10^{-14}$ | 1.1411 | 0.52843 | $1.3707 \cdot 10^{-7}$ | 1.5719 |
| slower w/o | 0.90445 | $2.5857 \cdot 10^{-16}$ | 1.0471 | 0.58633 | $6.7751 \cdot 10^{-9}$ | 1.2014 |

For high confidence of also small probabilities high precision arithmetic was used: $10^{-30}$ for computing the Poisson probabilities and matrix exponentials and their integrals in the synchronous approximations and $10^{-20}$ for solving the linear system. In the faster case the state space consists of $Kn+2$ states, where $n$ is the number of phases. The computations were done on a laptop with a single core Intel i7-1260P CPU 2.1 GHz with 32 GB RAM. Without expansion and with an expansion of two phases a run time of several 10 ms is required, with 10 phases about 1.5 s, and with 100 phases we get long run times of about 3 h, this is independent of the precision. This run time is however caused by Mathematica and with a more efficient implementation we would expect much smaller run times for 1002 states. Note that for the Erlang distribution with 100 phases a solution was even impossible with Mathematica. It must also be taken into account that the synchronized expansion leads to a fill-in and thus potentially to a higher computational effort for the same number of phases compared to the case when both servers are approximated by an Erlang distribution.

It can be seen that the faster and slower approximations without expansion already give very good approximations close to those with Erlang distributions with 100 phases or CME distributions with an order of 15. If the gap between both approximations is small enough for the modeling question, no further costly approximation by using any kind of expansion is necessary. It can also be seen that the system gets consistently slower from a faster expansion with two phases to a slower expansion with two phases and then to the slower approximation with additional states ("slower 1"). With 100 phases we get the slowest faster and the fastest slower approximation, as expected. The approximation quality of the slower approximations based on preemptions without expansion "slower w/o" is always better than that of the slower approximations based on additional states without expansion "slower 1" and sometimes even better than that of slower approximations with an expansion by few phases. Furthermore, the synchronous expansions with few phases already lead to closer approximations than compared with the Erlang distribution, this is especially significant for small values like $P(full)$. Furthermore, the Erlang approximations are sometimes not between the bounds from the synchronous expansions with the same number of phases and even the Erlang approximation with 100 phases is sometimes not inside the bounds. For other utilizations the results are similar. The approximation by CME distributions are within the bounds of the synchronous approximations, with one exception: for CME with order 15 and for $\lambda = 0.5$ the value of $P(full)$ is slightly outside the bounds of the synchronous approximations with 100 phases, which might be explained by remaining numerical errors; long running discrete-event simulations suggest results within the given bounds.

The results also show the impressive approximation quality of CME distributions. However, the linear system has positive and negative coefficients, which must be taken into account for the numerical solution. With the matrix-geometric formulation, a linear system of the size $\approx 2n^2$ must be solved, where $n$ is the order of the CME, which works here. When the CMEs would be used for the expansion of the state space without matrix-geometric solution, a larger linear system with negative entries results, which is numerically more difficult, especially when small probabilities are involved.

The results for $P(full)$ in case of light load ($\lambda = 0.1$) are also visualized in Fig. 7. For the slower approximation with one phase two values are given: the lower one corresponds to "slower w/o" (case 2), the upper to "slower 1" (case 1) showing clearly the better approximation quality by using preemptions. It also shows that the approximations based on Erlang distributions can be orders of magnitude off, whereas the approximations based on CME distributions are within the bounds of the synchronous approximations.

We have also tested with a uniform distribution from $0.5\tau$ to $1.5\tau$ (with $\lambda = 0.5, K = 10$) without expansion and obtained plausible bounds (faster: $P(idle) = 0.64260$, $P(full) = 7.3683 \cdot 10^{-9}$, $R = 0.92479$, slower: $P(idle) = 0.59132$, $P(full) = 2.5725 \cdot 10^{-8}$, $R = 1.1953$); it just needs a generalization of the integration of the matrix exponential and its integration as described in many sources, see e.g., [9]. Bernstein phase type distributions [11] with an order of up to 50 show the same tendency but are slightly outside the bounds for $P(full)$ whereas long running discrete-event simulations suggest results within the given bounds.

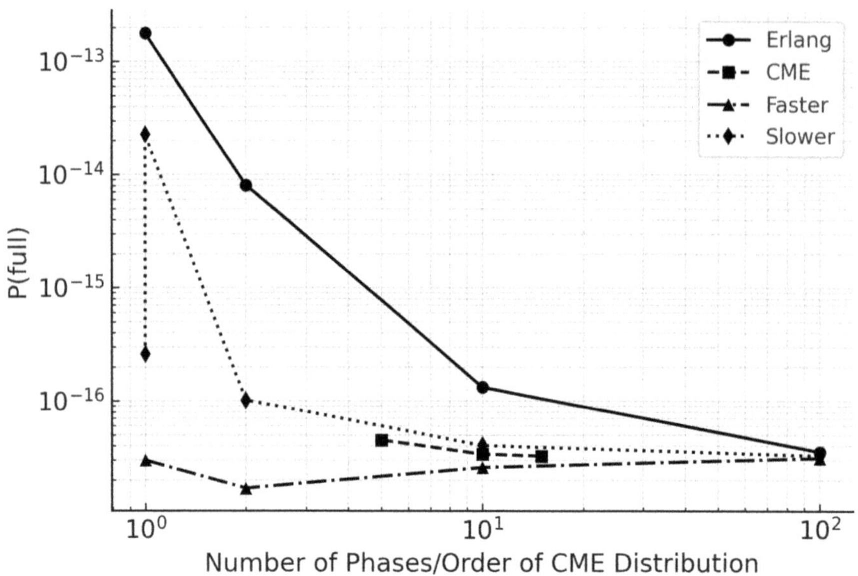

**Fig. 7.** Numerical results for $P(full)$ for $\lambda = 0.1$ and various kinds of approximation depending on the number of phases or the order of the CME distribution.

## 5   Conclusions

Synchronous approximations without and with phase expansions to get upper and lower bounds have been demonstrated by means of the $M/D/2/K$ system and with a Mathematica prototype. The bounds provide information whether the approximation is sufficient for the modeling task, this is not known from traditional phase-type expansions. The proper use of preemptions has been the key for a good slower approximation without phase expansion and was also needed for faster and slower approximations with phase expansions, in order to restart phases and keep the synchronization. Good results could be shown already for approximations without phase expansion, meaning that costly expansions of any kind can be avoided if the gap is small enough for the modeling task.

Many questions remain: the computational effort, the numerical stability (e.g., of Poisson probability and matrix exponential computation) and needed precision must be evaluated by means of a more advanced implementation (e.g., taking sparse data structures into account). Since synchronous phase expansions lead to a fill-in of sparse matrices, according to the matrix exponentiation, the tradeoff between numerical effort and accuracy needs to be investigated more thoroughly, especially in comparison with phase type and CME distributions. The approach of synchronous approximation should also be tested for other topologies as well, e.g., tandem queues, reliability models, or synchronized processes. The use for general distributions needs to be validated as well, e.g., by comparison with phase-type distributions. In case of non-identical concurrent non-Markovian activities synchronous phase expansion seems to be unavoidable, clear criteria to do this in the deterministic case can be formulated. Synchronous phase expansions for general distributions can be composed by convolution of identically distributed phases, this would be for instance possible for distributions closed under convolution, just few distributions do fulfill this criterion. It must be checked under which conditions the approach can be generalized and a general formulation how the state space is constructed without and with expansions is needed. The approach can be used for approximate transient analysis as well.

**Acknowledgments.** The author would like to thank Miklos Telek for his support with CMEs, with Bernstein phase type distributions, and formulating the analysis of the $M/D/2/K$ queuing system as a matrix-geometric problem with Mathematica BuTools. The author would also like to thank Peter Bazan for LaTeX support and drawing figures.

# References


1. Aldous, D., Shepp, L.: The least variable phase type distribution is Erlang. Commun. Stat.: Ser. Stochastic Models **3**(3), 467–473 (1987)
2. Biagi, M., Vicario, E., German, R.: Extending the steady state analysis of hierarchical semi-Markov processes with parallel regions. In: European Workshop on Performance Engineering, pp. 62–77 (2018)
3. Carnevali, L., German, R., Santoni, F., Vicario, E.: Compositional analysis of hierarchical UML statecharts. IEEE Trans. Soft. Eng. **48**(12), 4762–4788 (2021)
4. Çinlar., E.: Introduction to Stochastic Processes. Prentice Hall, Englewood Cliffs (1975)
5. Choi, H., Kulkarni, V.G., Trivedi, K.S.: Markov regenerative stochastic Petri nets. Perform. Eval. **20**, 337–357 (1994)
6. Ciardo, G.: Discrete-time Markovian stochastic Petri nets. In: Stewart, W. (ed.) Numerical Solution of Markov Chains, pp. 339–358. Raleigh (1995)
7. Cox, D.R.: The analysis of non-Markov stochastic processes by the inclusion of supplementary variables. Proc. Camb. Phil. Soc. (Math. and Phys. Sci.) **51**, 433–441 (1955)
8. Cox, D.R.: The use of complex probabilities in the theory of stochastic processes. Proc. Camb. Phil. Soc. (Math. and Phys. Sci.) **51**, 313–319 (1955)
9. German, R.: Performance Analysis of Communication Systems with Non-Markovian Stochastic Petri Nets. Wiley, Hoboken (2000)
10. Homm, D., German, R.: Analysis of hierarchical semi-markov processes with parallel regions. In: Remke, A., Haverkort, B.R. (eds.) MMB&DFT 2016. LNCS, vol. 9629, pp. 92–106. Springer, Cham (2016). https://doi.org/10.1007/978-3-319-31559-1_9
11. Horváth, A., Horváth, I., Paolieri, M., Telek, M., Vicario, E.: Approximation of cumulative distribution functions by Bernstein phase-type distributions. Perform. Eval. **168** (2025)
12. Horváth, A., Telek, M.: Phase Type Distributions: Theory and Application. Wiley, Hoboken (2024)

13. Horváth, G., Horváth, I., Telek, M.: High order concentrated matrix-exponential distributions. Stoch. Model. **32**(2), 176–192 (2019)
14. Lindemann, C.: Performance Modeling with Deterministic and Stochastic Petri Net. Wiley, Hoboken (1998)
15. Marsan, M.A., Balbo, G., Conte, G., Donatelli, S., Franceschinis, G.: Modelling with Generalized Stochastic Petri Nets. Wiley, Hoboken (1995)
16. Marsan, M.A., Chiola, G.: On Petri nets with deterministic and exponentially distributed firing times. In: Rozenberg, G. (ed.) Advances in Petri Nets 1987, pp. 132—-145 (1987)
17. Müller, A., Stoyan, D.: Comparison Methods for Stochastic Models and Risks. Wiley, Hoboken (2002)
18. Vicario, E., Sassoli, L., Carnevali, L.: Using stochastic state classes in quantitative evaluation of dense-time reactive systems. IEEE Trans. Softw. Eng. **35**(5), 703–719 (2009)

# The Pyramis Library: Efficient Numerical Evaluation of Hierarchical UML Statecharts Applied to Stochastic Workflows


Laura Carnevali[1], Reinhard German[2], Leonardo Montecchi[3],
Leonardo Scommegna[1(✉)], and Enrico Vicario[1]

[1] Information Engineering Department, University of Florence, Florence, Italy
{laura.carnevali,leonardo.scommegna,enrico.vicario}@unifi.it
[2] Computer Science Department, Friedrich-Alexander-Universität
Erlangen-Nürnberg, Erlangen, Germany
reinhard.german@fau.de
[3] Computer Science Department, Norwegian University of Science and Technology,
Trondheim, Norway
leonardo.montecchi@ntnu.no



**Abstract.** Pyramis is a library for quantitative evaluation of hierarchical UML statecharts with non-Markovian stochastic timing and probabilistic choices. It implements an efficient numerical approach for transient analysis until absorption and steady-state analysis, separately evaluating the Semi-Markov Process (SMP) of each model component. As Pyramis facilitates code reusability, maintainability, and extensibility, it has been easily integrated with the FaultFlow library for dependability analysis of component-based systems, supporting efficient quantitative evaluation of stochastic static fault trees without repeated events.

In this paper, we use Pyramis for quantitative evaluation of workflows where activities have non-Markovian stochastic duration and where precedence constraints define a Directed Acyclic Graph (DAG). Workflows have a Service Level Objective (SLO) on their end-to-end (E2E) response time distribution at low workloads of requests. Pyramis efficiently derives the workflow E2E response time distribution, yielding a stochastic upper bound for topologies with non-well-nested precedence DAGs. In our experiments, we consider a workflow with topology derived from a real benchmark and execution times obtained from a known dataset. We report results for workflow variants obtained by increasing the number of sequential, concurrent, or alternative activities of workflow patterns. Results are promising in terms of tradeoff between accuracy and complexity.

**Keywords:** Statecharts · workflows with stochastic durations · non-Markovian distributions · semi-Markov processes · software tools and libraries

# 1    Introduction

*Motivation.* Quantitative evaluation of non-Markovian models has the potential to effectively support performance and dependability engineering of complex time-critical systems. To this end, the adopted formalism should guarantee not only ease of model construction, but also harmonization with consolidated industrial practice [25, 38], and analyzability of the underlying stochastic process. In addition, a software tool should support stochastic modeling and analysis. Addressing these intertwined aspects is challenging and fundamental to success.

*Related Works.* One of the solutions in the literature is Model Driven Engineering (MDE) [16, 42], which pursues this goal by automatically deriving stochastic models from semi-formal artifacts [4, 28, 36]. These artifacts are typically specified by diagrams of the Unified Modeling Language (UML), annotated by the profiles for Modeling and Analysis of Real-Time Embedded systems (MARTE) [7, 33] and for Dependability Analysis and Modeling (DAM) [6], or by diagrams of the Systems Modeling Language (SysML) [34].

Most of the MDE literature translates semi-formal artifacts into Markovian models [10, 24, 27, 31, 44, 48], using consolidated efficient solutions [2, 3, 22, 39] while limiting model expressivity with exponential timers only, hindering representation of synchronous events (e.g., timeouts) and history-dependent behaviors (e.g., aging). Few MDE approaches exploit non-Markovian models [5, 8, 26, 30, 45, 50], improving model expressivity while facing harder complexities in evaluating the underlying stochastic process [21]. To mitigate the issue, these methods either limit the number of concurrent non-Markovian timers, or consider classes of domain-specific models whose concurrency structure and stochastic timing guarantee feasibility of stochastic analysis [9, 13, 14, 18, 19, 23, 49]. The method of [18, 19] performs availability analysis of dynamic fault trees, using the model structure to compute stochastic bounds on the time to failure distribution and evaluate maintenance policies. The work in [49] derives the execution time distribution of a workflow with precedence constraints defined by a Directed Acyclic Graph (DAG), fitting general distributions with phase-type distributions, and combining numerical evaluation of subgraphs having series or parallel topology with probabilistic model checking of the other subgraphs. The compositional approach of [14] yields a stochastic upper bound on the end-to-end (E2E) response time distribution of workflows with DAG precedence constraints and non-Markovian service durations, a relevant class of stochastic models in performance engineering [41]. The work in [23] has suggested to address stochastic modeling and analysis of a hierarchical extension of UML statecharts with non-Markovian stochastic timing and probabilistic choices, termed Hierarchical Semi-Markov Process with parallel regions (HSMP). It has been extended in [9] by the time advance mechanism known from stochastic state classes [47] to take exits in parallel regions with different time origins correctly into account. In [13] it was put on a formal semantical basis, extended further, and implemented in the Pyramis library. The HSMP formalism provides ease of representation, while reducing the model expressiveness with fairly lax restrictions that enable separate analysis of the Semi-Markov Process (SMP) underlying each model component, guaranteeing efficient evaluation of steady-state and transient behavior (until absorption).

*Contribution.* Pyramis [13] is a Java library implementing the approach of [13,23], using consolidated design patterns [20] and MDE principles to support code reusability, maintainability, extensibility, and integrability. It is available open source under the AGPLv3 licence [37]. In this paper, we use Pyramis for quantitative evaluation of workflows composing activities with DAG precedence constraints and non-Markovian durations. Workflows have a Service Level Objective (SLO) on their E2E response time distribution at low workloads. We translate the UML activity diagram of a workflow into an HSMP, disregarding dependencies in non-well-nested precedence DAGs, while efficiently deriving a stochastic upper bound on the workflow E2E response time distribution. We analyze a set of workflows with topology derived from a real benchmark and execution times obtained from a dataset of the literature, achieving a trade-off between accuracy of evaluation and complexity of computation. A package enabling replication of the experimental results is publicly available under the AGPLv3 licence at https://github.com/oris-tool/pyramis-evaluation-rep-pkg.

*Structure.* In the following, we discuss the Pyramis metamodel (Sect. 2) and we use it to model workflows with stochastic durations (Sect. 3). Then, we illustrate experimental results (Sect. 4) and we draw conclusions (Sect. 5).

## 2   An Overview of the Pyramis Library

Pyramis is a Java library for quantitative modeling and evaluation of HSMPs [9, 13,23], an extension of UML statecharts capturing the concepts of concurrency, hierarchy, non-Markovian stochastic timing, and probabilistic choices.

*HSMP syntax.* Figure 1 shows the Pyramis metamodel using UML class diagram notation, and Fig. 3b shows the HSMP of a workflow (we discuss in Sect. 4 how to derive the HSMP of Fig. 3b from the UML activity diagram of Fig. 3a).

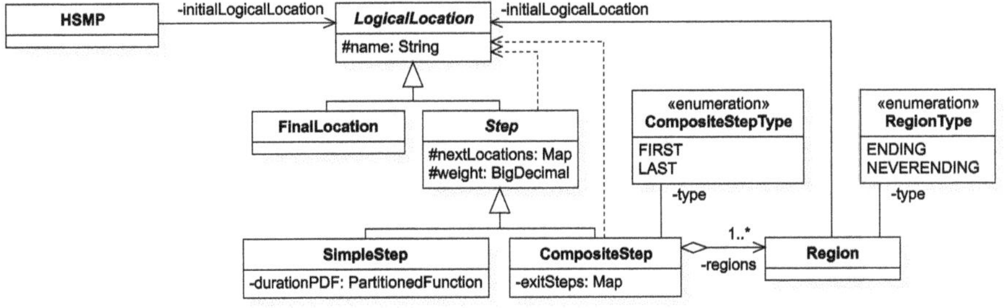

**Fig. 1.** Metamodel of the Pyramis library, using UML class diagram notation.

Specifically, an HSMP (modeled by class HSMP in Fig. 1) contains locations (modeled by class LogicalLocation in Fig. 1) which can be either *steps* (modeled by class Step in Fig. 1 and depicted as rounded boxes in Fig. 3b), i.e., actions with stochastic duration, or *final locations* (modeled by class FinalLocation in Fig. 1), i.e., terminated actions. In turn, steps can be either *simple* (modeled by class SimpleStep in Fig. 1), i.e., atomic actions having a duration distribution, or *composite* (modeled by class CompositeStep in Fig. 1), i.e., composed actions made of *concurrent regions* (modeled by class Region in Fig. 1 and separated by dashed lines in Fig. 3b). Regions can be of type either ENDING (i.e., terminating in a final location) or NEVERENDING (i.e., not terminating). We collect regions in composite step of type either LAST (i.e., terminating as soon as all its regions have terminated, with final locations depicted in Fig. 3b as unfilled circles on the border of the region, with a smaller black-filled circle inside) or FIRST (i.e., terminating as soon as any of its regions terminates, with final locations depicted as unfilled circles on the border of the region, with an Xshaped cross inside)[1].

Composition of regions and steps yields a hierarchy of HSMPs where each composite step is defined by an HSMP at the next lower level (e.g., composite step B in Fig. 3b is itself defined by an HSMP), the top-level HSMP contains a single region (i.e., the top-level region of Fig. 3b contains a single region composing composite step AtoM and simple steps N and O in sequence), and non-top-level HSMPs contain at least one region of type ENDING (e.g., the regions of all the composite steps in Fig. 3b are of type ENDING).

*HSMP Semantics.* The state of an HSMP consists of: an active location for each region of the composite step modeling the HSPM; a time to the next event (TNE) for each active step (modeling the time to the completion of the step); and, a state for each active composite step. As an example, we consider the initial state of the HSMP of Fig. 3b, where all the simple steps contained in composite steps $A'$, $A''$, and B are active and sample a TNE from their duration distribution. After each event (i.e., step completion), the TNE of each active step is reduced by the TNE of the completed step. If all the steps of $A'$ terminate before all the steps of $A''$ and B have terminated, then $C'$ becomes the newly active step of region RACE′ and samples a TNE. Otherwise, if all the steps of $A''$ terminate before all the steps of $A'$ and B have terminated, then $C''$ becomes the newly active step of region RAC″ and samples a TNE; conversely (i.e., if all the steps of B terminate before all the steps of $A'$ and $A''$ have terminated), D becomes the newly active step of region RBD and samples a TNE. Then, the model execution proceeds event after event. Leveraging this semantics, the approach of [13] implemented by Pyramis separately derives the duration distribution of each composite step by separately evaluating the SMP of each of its regions. Then, the approach exploits these distributions to derive the steady-state probability of each step.

---

[1] Composite steps of type FIRST are not shown in the HSMP model of Fig. 3b due to the fact that they are not used to model the considered class of workflows.

*Pyramis Features.* Pyramis is designed following consolidated design patterns [20] and MDE principles [16,42] to support code reusability, maintainability, and extensibility, and integration with other libraries. As evidence of this fact, Pyramis has been easily integrated with FaultFlow [11,12], a Java library for dependability evaluation of component-based systems, with focus on fault propagations within individual components and between different components. The main outcome of such integration was the automated translation of stochastic static fault trees without repeated events from the FaultFlow metamodel to the Pyramis metamodel, enabling the exploitation of the solution method implemented by Pyramis to derive the distribution of the time to the occurrence of any component failure (including the system failure) and the Birnbaum and Fussell-Vesely importance measures of faults. Notably, Pyramis efficiently affords the analysis of very complex SSFTs with large depth and hundreds of different faults.

## 3   Quantitative Modeling of Workflows Using Pyramis

We consider workflows composing actions with durations having non-Markovian distributions and with precedence constraints defining a DAG. Figure 2 shows the UML activity diagram of a workflow, extending the topology of the Sipth workflow of the Pegasus repository [1,17] to include all action composition patterns for illustration purposes. Specifically, a workflow consists of *elementary actions* modeling atomic operations (e.g., A1.1 and A1.2) and *structured actions* (labeled with stereotype <<structured>>) composing other (elementary or structured) actions (e.g., A and B). A workflow composes actions with different constructs: *sequence* (e.g., A1 models the sequential execution of A1.1 and A1.2); *fork-join* (e.g., A models the concurrent execution of A1 and A2); *split-merge* (e.g., M models the exclusive execution of either M1 or M2, both with probability 0.5); *irreducible DAG*, i.e., a DAG that is made of *individual* fork and join operators and cannot be modeled by sequence and fork-join constructs previously introduced[2] (e.g., TOP composing AC, BD, E, FtoM, and NO). A *non-well-nested* DAG includes irreducible DAG constructs, while a *well-nested* DAG does not.

By construction, HSMPs cannot represent a workflow that composes actions with irreducible DAG constructs. Rather, they can model a well-nested variant of the workflow where we replicate the actions belonging to different paths of irreducible DAG constructs (in a DAG backward visit, we replicate each action that is predecessor of more than one action or of a replicated action). Figure 3a shows the UML activity diagram of such variant of the workflow of Fig. 2, obtained by replicating AC (predecessor of E and FtoM) into AC' and AC''. Replication of predecessor actions guarantees that the E2E response time distribution $F$ of the

---

[2] Individual fork and join operators are different constructs than the fork-join one.

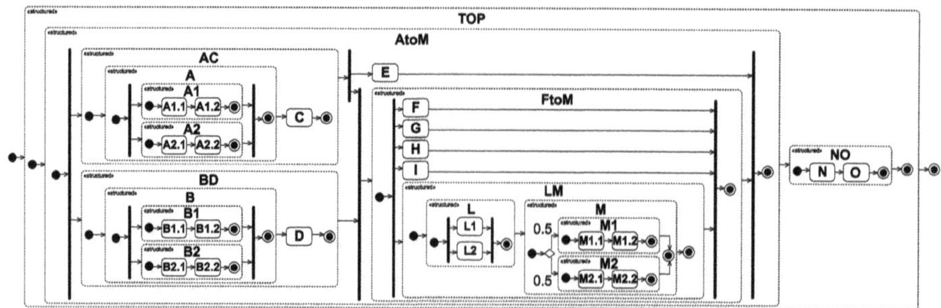

**Fig. 2.** UML activity diagram of a variant of the Sipht workflow of [1,17] (the name of a structured action indicates the names of the actions that it composes).

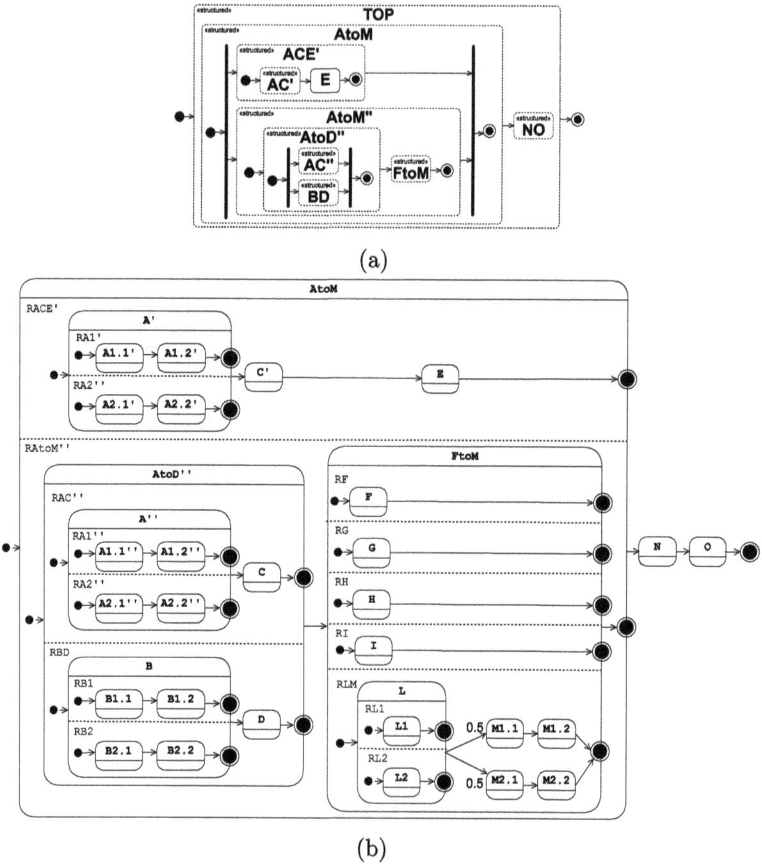

**Fig. 3.** (a) UML activity diagram of a variant of the workflow of Fig. 2, obtained by replicating AC into AC' and AC'' (the name of a structured action contains the symbol ' or the symbol '' depending on whether the structured action contains AC' or AC'', respectively).(b) HSMP of the workflow of Fig. 3b.

obtained workflow is a *stochastic upper bound* of the E2E response time distribution $G$ of the actual workflow (i.e., $F(t) \leq G(t) \; \forall t$), as proved in [14], e.g., in Fig. 4a, the E2E response time distributions of the workflow of Fig. 2 computed by Pyramis with different time tick (solid lines) are stochastic upper bounds of the ground truth distribution obtained by performing simulation(dashed line).

Then, we easily derive the HSMP of a workflow from the UML activity diagram of the mentioned variant of the workflow, by mapping: elementary actions into steps; sequence constructs into a sequence of steps, one predecessor of the other; fork-join constructs into composite steps of type LAST; and, split-merge constructs into steps with multiple successor steps. Figure 3b shows the HSMP of the workflow of Fig. 2 obtained from the UML activity diagram of Fig. 3a.

## 4   Quantitative Evaluation of Workflows Using Pyramis

In this section, first we discuss our experimental methodology (Sect. 4.1) and results (Sect. 4.2). Then, we provide some final remarks (Sect. 4.3) and we discuss the threats to validity (Sect. 4.4). We have performed the experiments on a single core of an Intel i7-1260P CPU 4.70 GHz with 32.0 GB RAM.

### 4.1   Experimental Methodology

*Workflow Topology.* We define variants of the workflow topology of Fig. 2 to stress relevant complexity factors of HSMP stochastic analysis:,the *sequencing degree* $s$ of a sequence construct (i.e., the number of sequential actions of the construct), the *concurrency degree* $c$ of a fork-joint construct (i.e., the number of concurrent actions of the construct), and the *split-merge degree* $m$ of a split-merge construct (i.e., the number of alternative actions of the construct). Specifically, we modify the topology of Fig. 2 in three different directions: $i$) we increase the sequencing degree $s$ of the structured actions A1, A2, B1, and B2 by considering $s \in \{2, 4, 8\}$ ($s = 2$ is the case of the topology of Fig. 2); $ii$) we increase the concurrency degree $c$ of the structured actions A and B, by considering $c \in \{2, 4, 8\}$ ($c = 2$ is the case of the topology of Fig. 2); $iii$) we increase the split-merge degree $m$ of the structured action M by considering $m \in \{2, 4, 8\}$ ($m = 2$ is the case of the topology of Fig. 2). One dimension is modified at a time, keeping the others unchanged, yielding 7 different workflow topologies. Note that each considered workflow is non-well-nested as it includes the irreducible DAG TOP.

*Workflow Stochastic Durations.* For each of the 7 topologies, we define a duration distribution for each elementary action by approximating a histogram of samples derived from the WS-Dream dataset [51], which collects response times of many web services. Specifically, we consider 100 web services and, for each of them, we derive a histogram by collecting and regularizing the related samples according to the inter-quartile range rule [46]. Then, for each elementary action of our workflow, we randomly select a histogram and we approximate it with the distribution of a shifted truncated exponential random variable fitting its mean value and its coefficient of variation [15], i.e., a random variable having support $[a, b]$, probability density function $h(t) := \lambda\, e^{-\lambda t}\, e^{a\lambda}/(1 - e^{(a-b)\lambda})$ for some rate $\lambda > 0$, and probability distribution function $H(t) := (1 - e^{a\lambda} e^{-\lambda t})/(1 - e^{(a-b)\lambda})$.

*Evaluated Reward.* For each workflow, we translate the UML activity diagram of the workflow into an HSMP model (as illustrated in Sect. 3). Then, we perform transient analysis until absorption of the obtained HSMP, deriving the Cumulative Distribution Function (CDF) $F$ of the duration of the top-level composite step of the HSMP. In particular, note that $F$ comprises a stochastic upper bound on the workflow E2E response time CDF, due to the fact that the workflow is non-well-nested (as discussed at the end of Sect. 3).

For each of the 7 considered workflows, we repeat the analysis by Pyramis with time tick 0.4, 0.2, and 0.1, using time limit equal to 25 for each model.

*Ground Truth Derivation.* For each workflow, we evaluate a ground-truth E2E response time CDF $G$ by performing 1 million simulation runs of the Stochastic Time Petri Net (STPN) that models the workflow, by leveraging the SIRIO library [43] of the ORIS tool [35]. Specifically, for each workflow, we use time tick 0.1 and the same limit equal to 25 used in the analysis by Pyramis. Then, we derive the Probability Density Functions (PDFs) $f$ and $g$ from the CDFs $F$ and $G$, respectively, and we evaluate the accuracy of $f$ by using the Jensen-Shannon (JS) divergence $D_{\mathrm{JS}}(f \,||\, g)$ of $f$ from $g$ [29,32], which is defined as:

$$D_{\mathrm{JS}}(f \,||\, g) = \frac{1}{2} D_{\mathrm{KL}}(f \,||\, z) + \frac{1}{2} D_{\mathrm{KL}}(g \,||\, z), \tag{1}$$

where $z(t) = \frac{1}{2}(f(t) + g(t))\ \forall\, t \in \Omega$ averages $f$ and $g$, $\Omega$ is a set of equidistant time points covering the supports of $f$ and $g$, and $D_{\mathrm{KL}}(f \,||\, z)$ is the Kullback-Leibler (KL) divergence of $f$ from $z$. In turn, $D_{\mathrm{KL}}(f \,||\, z)$ is defined as:

$$D_{\mathrm{KL}}(f \,||\, z) = \sum_{t \in \Omega} f(t) \cdot \log\left(\frac{z(t)}{f(t)}\right). \tag{2}$$

*Baseline Simulation.* For each workflow, using time tick equal to 0.1, we also perform a number of simulation runs that require a total computation time equal to the running time of the analysis performed by Pyramis with time tick 0.1, evaluating the corresponding workflow E2E response time CDF $H$ and PDF $h$. Then, we evaluate the accuracy of the obtained simulative baseline through the JS divergence $D_{\mathrm{JS}}(h, g)$ of $h$ from the ground-truth E2E response time PDF $g$.

### 4.2   Experimental Results

Table 1 shows the running times for all experiments, including the time taken for each considered workflow to generate the ground truth through simulation. In particular, the running times of each analysis by Pyramis are obtained by averaging the running times of 100 repetitions of that analysis. Moreover, for each workflow, Table 1 also shows the JS divergence obtained by comparing both the Pyramis analysis and the baseline simulation against the ground truth.

**Table 1.** For each workflow, the running times of the analysis (time tick in $\{0.4, 0.2, 0.1\}$) and ground truth derivation (time tick 0.1), and the JS divergence of the analysis and simulative baseline (time tick 0.1) from the ground truth.

| workflow variant | time tick | analysis time (ms) | GT time (min) | $D_{\mathrm{JS}}\,(f \,\|\, g)$ | $D_{\mathrm{JS}}\,(h \,\|\, g)$ |
|---|---|---|---|---|---|
| $s = 2,\, c = 2,\, m = 2$ | 0.4 | 4.87 | | $3.908 \times 10^{-3}$ | |
| | 0.2 | 5.37 | | $1.065 \times 10^{-3}$ | |
| | 0.1 | 15.62 | 7.07 | $3.247 \times 10^{-4}$ | $2.524 \times 10^{-4}$ |
| $s = 4\ (c = 2,\, m = 2)$ | 0.4 | 2.60 | | $4.514 \times 10^{-3}$ | |
| | 0.2 | 7.24 | | $1.166 \times 10^{-3}$ | |
| | 0.1 | 25.48 | 9.60 | $3.146 \times 10^{-4}$ | $2.252 \times 10^{-4}$ |
| $s = 8\ (c = 2,\, m = 2)$ | 0.4 | 7.08 | | $1.139 \times 10^{-2}$ | |
| | 0.2 | 23.20 | | $4.275 \times 10^{-3}$ | |
| | 0.1 | 83.59 | 14.47 | $2.013 \times 10^{-3}$ | $1.803 \times 10^{-4}$ |
| $c = 4\ (s = 2,\, m = 2)$ | 0.4 | 2.81 | | $3.990 \times 10^{-3}$ | |
| | 0.2 | 5.79 | | $1.090 \times 10^{-3}$ | |
| | 0.1 | 15.44 | 18.93 | $3.360 \times 10^{-4}$ | $1.152 \times 10^{-4}$ |
| $c = 8\ (s = 2,\, m = 2)$ | 0.4 | 3.19 | | $3.586 \times 10^{-3}$ | |
| | 0.2 | 8.37 | | $9.105 \times 10^{-4}$ | |
| | 0.1 | 27.10 | 20.47 | $2.541 \times 10^{-4}$ | $3.059 \times 10^{-4}$ |
| $m = 4\ (s = 2,\, c = 2)$ | 0.4 | 1.98 | | $3.986 \times 10^{-3}$ | |
| | 0.2 | 5.64 | | $1.085 \times 10^{-3}$ | |
| | 0.1 | 20.60 | 7.51 | $3.340 \times 10^{-4}$ | $2.536 \times 10^{-4}$ |
| $m = 8\ (s = 2,\, c = 2)$ | 0.4 | 3.97 | | $3.930 \times 10^{-3}$ | |
| | 0.2 | 12.61 | | $1.079 \times 10^{-3}$ | |
| | 0.1 | 46.39 | 8.15 | $3.318 \times 10^{-4}$ | $4.299 \times 10^{-4}$ |

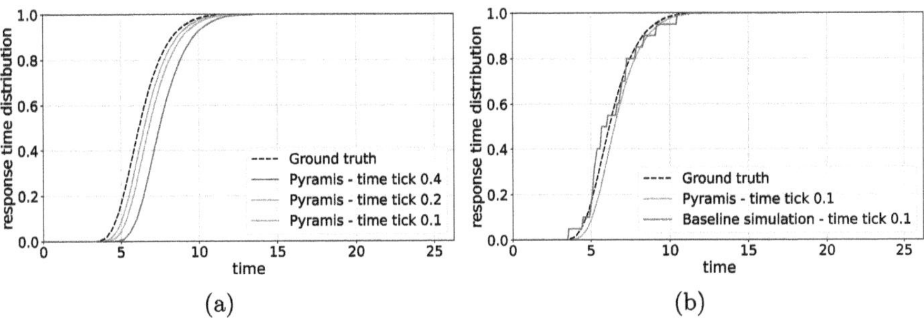


**Fig. 4.** E2E response time distribution of the workflow of Fig. 2.

Table 1 shows a marked time gap between the analyses run with Pyramis and the ground truth generated through simulation, which in fact required one million runs. The Pyramis analyses never exceeded an average of 100 ms, whereas producing the ground truth always took more than 7 min. Table 1 also shows that, for each workflow, both the running time and the result accuracy achieved by Pyramis depend on the time tick: a coarser tick actually yields faster results but increases the JS divergence $D_{\mathrm{JS}}(f \parallel g)$ from the ground truth.

The analysis results for the workflow in Fig. 2 (i.e., $s = 2$, $c = 2$, $m = 2$) are presented in numbers in Table 1 and graphically in Fig. 4. Specifically, Fig. 4a plots the E2E response time CDF provided by the ground truth against the E2E response time CDFs calculated by the Pyramis analysis using time tick 0.4, 0.2, and 0.1. As already noted, using a finer time tick leads to a lower JS divergence from the ground truth. Graphically, this is reflected by the analysis CDF approaching the ground truth CDF from the right. Also note that the accuracy improvement decreases with the time tick, as the analysis CDF remains a stochastic upper bound of the ground-truth CDF, as in Fig. 4a. Figure 4b compares the analysis CDF obtained with time tick 0.1, the ground-truth CDF, and the CDF computed by a simulative baseline (20 runs) with time tick 0.1 having the same computation time of the Pyramis analysis. Notably, simulation achieves nearly the same JS distance from the ground truth as Pyramis (see Table 1), while not providing a stochastic upper bound of the ground-truth CDF.

Figure 5 shows the graphical analysis results for the variants of the workflow in Fig. 2 with different sequencing degree (i.e., $s = 4$, $s = 8$). Figures 5a and 5c compare the analysis CDF obtained with different time tick with the ground-truth CDF for $s = 4$ and $s = 8$, respectively, showing that the analysis

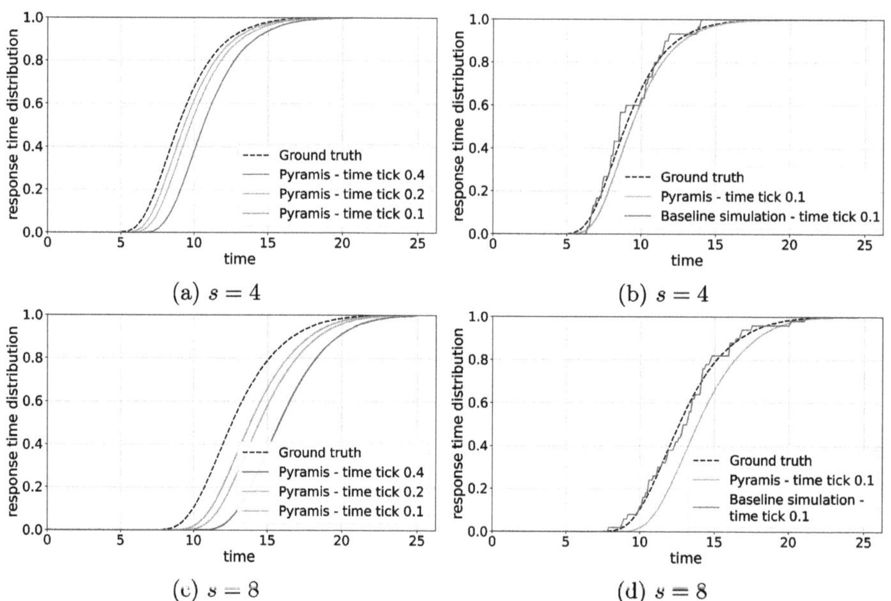

**Fig. 5.** E2E response time CDF of variants of the workflow of Fig. 2, obtained by varying the sequencing degree $s$ of the structured actions A1, A2, B1, and B2.

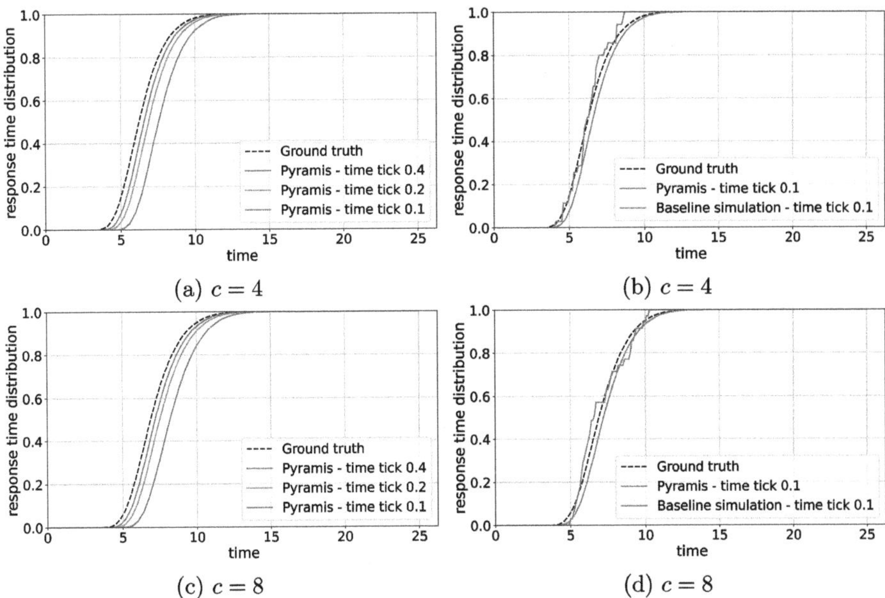

**Fig. 6.** E2E response time CDF of variants of the workflow of Fig. 2, obtained by varying the concurrency degree $c$ of the structured actions **A** and **B**.

CDF becomes closer to the ground-truth CDF as the time tick decreases, with improvement decreasing with the time tick. Figures 5b and 5d compare the analysis CDF with time tick 0.1 and the ground-truth CDF with the CDF provided by a baseline simulation having the same running time of the analysis for $s = 4$ (30 runs) and $s = 8$ (50 runs), respectively. The results again show that the simulative baseline yields JS divergence from the ground truth in the same order of magnitude as that of the analysis, though not guaranteeing that the obtained CDF is a stochastic upper bound of the ground-truth CDF.

Figure 6 presents the results for the workflow variants with different concurrency degree (i.e., $c = 4$, $c = 8$). In particular, we obtain the simulative baselines with the same running time as that of the analysis by performing 35 runs both for $c = 4$ (Fig. 6b) and for $c = 8$ (Fig. 6d), given the similar analysis running time in the two cases (i.e., approximately 12 ms, as reported in Table 1). Finally, Fig. 7 shows the plots related to the workflow variants with different split-merge degree (i.e., $m = 4$ and $m = 8$). In particular, we obtain the simulative baselines with the same running time as that of the analysis by performing 30 runs for $m = 4$ (Fig. 7b) and 50 for $m = 8$ (Fig. 7d). Overall, the results in Figs. 6 and 7 show the same patterns observed for the workflow variants in Figs. 4 and 5. For the analysis, a finer time tick yields a smaller JS divergence, maintaining stochastic ordering with respect to the ground truth. In contrast, the simulative baselines do not. Note that some topology variations affect accuracy more than others, e.g., for $s = 8$, the JS divergence remains in the order of $10^{-3}$ even with time tick 0.1, which is due to the fact that the E2E response time CDF of a sequence construct is obtained by performing the convolution of the E2E response time PDFs of its sequential actions, which is more prone to numerical errors than the algebraic operations performed for the fork-join and split-merge constructs [13].

## 4.3   Remarks

We examined the efficiency and accuracy of the analysis by Pyramis in evaluating the E2E response time CDF of complex workflows. Results show that the analysis offers a clear computational advantage over simulation, with execution times in the order of ms for the analysis and of min for the ground-truth simulation. Moreover, the analysis accuracy improves as the time tick decreases, being in the same order of magnitude of that achieved by a simulative baseline with the same time tick. Notably, unlike Pyramis, a simulative baseline cannot guarantee stochastic ordering of the obtained CDF with respect to the ground-truth CDF.

Analysis and simulation are not interchangeable but rather complementary. Simulation is preferable if the running time is not critical and stochastic ordering of the E2E response time CDF is not needed. In contrast, our analysis method is better suited to time-bounded scenarios (e.g., reactive systems) where stochastic upper bounds are necessary (e.g., performance-constrained environments).

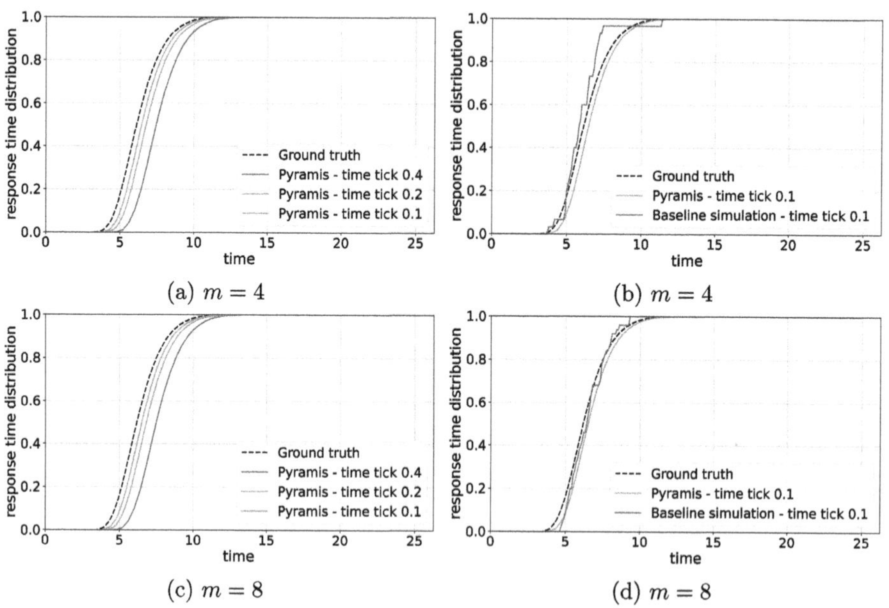

**Fig. 7.** E2E response time CDF of variants of the workflow of Fig. 2, obtained by varying the split-merge degree $m$ of the structured action M.

## 4.4   Threats to Validity

We discuss the threats to validity by following the classification provided in [40].

*Construct Validity.* To mitigate inadequate operationalization, a key threat to construct validity, we evaluated our analysis through a multi-faceted approach. Specifically, we compared the time taken to evaluate the workflow E2E response time CDF against that of a simulative baseline, and we assessed the achieved accuracy in terms of the JS divergence from a simulative ground truth, showing also experimentally that our CDF is a stochastic upper bound of the actual one.

*Internal Validity.* The way in which we design and parameterize the experimental could invalidate the results and lead to erroneous conclusions. To mitigate this threat to internal validity, we based our experiments on a workflow topology described in the literature [1,17] and, for parameterization, we used distributions used derived from a dataset of actual processing times observed in practice [51].

*External Validity.* The results of our experiments may be biased by the workflow topology. To mitigate this threat to external validity, we chose a realistic topology inspired by a workflow of the literature, and we built variants by modifying the complexity of some of its sequence, fork-join, and split-merge constructs.

*Reliability.* For the independent reproduction and verification of our results, we have made a replication package available open-source online, including the workflow topologies and the parameterizing distributions.

## 5   Conclusions

We have used Pyramis, a library for efficient numerical evaluation of hierarchical UML statecharts, to compute a stochastic upper bound on the E2E response time distribution of stochastic workflows. We plan to integrate Pyramis with Eulero, a library implementing the method of [14] to achieve the same goal. The work in [14] limits the number of simplifications performed to manage dependencies in non-well-nested precedence DAGs, achieving better evaluation accuracy than Pyramis while incurring greater computational complexity. To compare Pyramis with Eulero, we could leverage onsolidated MDE principles to let Pyramis to accept workflows in the Eulero format. Pyramis could be used also to analyze subworkflows and support decisions taken by the Eulero compositional solution (e.g., identifying the most complex subworkflow, to be analyzed separately).

## References


1. Pegasus workflow repository, https://pegasus.isi.edu/workflow_gallery
2. Ajmone Marsan, M., Conte, G., Balbo, G.: A Class of Generalized Stochastic Petri Nets for the Performance Evaluation of Multiprocessor Systems. ACM Trans. Comput. Syst. **2**(2), 93–122 (1984). https://doi.org/10.1145/190.191

3. Baier, C., Katoen, J.P.: Principles of Model Checking. MIT Press (2008)
4. Balsamo, S., Di Marco, A., Inverardi, P., Simeoni, M.: Model-based performance prediction in software development: a survey. IEEE Trans. SW Eng. **30**(5), 295–310 (2004). https://doi.org/10.1109/TSE.2004.9
5. Bernardi, S., Campos, J., Merseguer, J.: Timing-failure risk assessment of UML design using Time Petri Net bound techniques. IEEE Trans. on Industrial Inf. **7**(1), 90–104 (2011). https://doi.org/10.1109/TII.2010.2098415
6. Bernardi, S., Merseguer, J., Petriu, D.C.: A dependability profile within MARTE. Soft. & Sys. Model. **10**(3), 313–336 (2011)
7. Bernardi, S., Merseguer, J., Petriu, D.C.: Model-Driven Dependability Assessment of Software Systems. Springer Berlin, Heidelberg (2013). https://doi.org/10.1007/978-3-642-39512-3
8. Biagi, M., Carnevali, L., Tarani, F., Vicario, E.: Model-based quantitative evaluation of repair procedures in gas distribution networks. ACM Trans. Cyber-Phys. Sys. **3**(2), 19:1–19:26 (2019)
9. Biagi, M., Vicario, E., German, R.: Extending the steady state analysis of hierarchical semi-Markov processes with parallel regions. In: European Workshop on Perf. Eng. pp. 62–77 (2018)
10. Bondavalli, A., Dal Cin, M., Latella, D., Majzik, I., Pataricza, A., Savoia, G.: Dependability analysis in the early phases of UML-based system design. Comp. Syst. Sci. Eng. **16**(5), 265–275 (2001)
11. Carnevali, L., Cerboni, S., Montecchi, L., Vicario, E.: FaultFlow: an MDE Library for Dependability Evaluation of Component-Based Systems. IEEE Trans. Dependable Secure Comput. (2025)
12. Carnevali, L., Cerboni, S., Picano, B., Scommegna, L., Vicario, E.: An observation metamodel for dependability tools. In: European Dependable Computing Conf. (EDCC). pp. 169–172. IEEE (2024)
13. Carnevali, L., German, R., Santoni, F., Vicario, E.: Compositional Analysis of Hierarchical UML Statecharts. IEEE Trans. on Soft. Eng. **48**(12), 4762–4788 (2021)
14. Carnevali, L., Paolieri, M., Reali, R., Vicario, E.: Compositional safe approximation of response time probability density function of complex workflows. ACM Trans. on Model. Comput. Simul. **33**(4) (2023)
15. Carnevali, L., Reali, R., Vicario, E.: Compositional evaluation of stochastic workflows for response time analysis of composite web services. In: ACM/SPEC International Conference on Performance Engineering, pp. 177–188 (2021)
16. Da Silva, A.R.: Model-driven engineering: a survey supported by the unified conceptual model. Comp. Lang., Sys. & Struct. **43**, 139–155 (2015)
17. Deelman, E., Vahi, K., Rynge, M., Mayani, R., da Silva, R.F., Papadimitriou, G., Livny, M.: The evolution of the pegasus workflow management software. Comput. Sci. & Eng. **21**(4), 22–36 (2019)
18. Fourneau, J.M., Pekergin, N.: A Numerical Analysis of Dynamic Fault Trees Based on Stochastic Bounds. In: Campos, J., Haverkort, B.R. (eds.) QEST 2015. LNCS, vol. 9259, pp. 176–191. Springer, Cham (2015). https://doi.org/10.1007/978-3-319-22264-6_12
19. Fourneau, J., Pekergin, N.: Dynamic fault trees with rejuvenation: Numerical analysis and stochastic bounds. Electron. Notes Theor. Comput. Sci. **327**, 27–47 (2016)
20. Gamma, E., Helm, R., Johnson, R., Vlissides, J., Patterns, D.: Elements of reusable object-oriented software. Design Patterns (1995)
21. German, R.: Performance Analysis of Communication Systems with Non-Markovian Stochastic Petri Nets. John Wiley (2000)

22. Grassmann, W.: Transient solutions in Markovian queues: an algorithm for finding them and determining their waiting-time distributions. Eur. J. Oper. Res. **1**(6), 396–402 (1977)
23. Homm, D., German, R.: Analysis of hierarchical semi-Markov processes with parallel regions. In: Int. GI/ITG Conf. Measurement, Modelling, and Evaluation of Comp. Sys. and Dependability and Fault Tolerance. pp. 92–106. Springer (2016)
24. Huszerl, G., Majzik, I., Pataricza, A., Kosmidis, K., Dal Cin, M.: Quantitative analysis of UML statechart models of dependable systems. Comput. J. **45**(3), 260–277 (2002)
25. International Organization for Standardization: ISO 9241 - Ergonomic requirements for office work with visual display terminals (VDTs) (2000)
26. Jansen, D., Hermanns, H., Katoen, J.P.: A QoS-oriented extension of UML statecharts. In: Stevens, P., Whittle, J., Booch, G. (eds) International Conference on the Unified Modeling Language pp. 76–91. Springer (2003). https://doi.org/10.1007/978-3-540-45221-8_7
27. King, P., Pooley, R.: Derivation of Petri net performance models from UML specifications of communications software. In: Haverkort, B.R., Bohnenkamp, H.C., Smith, C.U. (eds.) International Conference on Modelling Techniques and Tools for Computer Performance Evaluation. pp. 262–276. Springer, Berlin, Heidelberg (2000). https://doi.org/10.1007/3-540-46429-8_19
28. Koziolek, H.: Performance evaluation of component-based software systems: A survey. Perf. Eval. **67**(8), 634–658 (2010). https://doi.org/10.1016/j.peva.2009.07.007
29. Lin, J.: Divergence measures based on the shannon entropy. IEEE Trans. Inf. theory **37**(1), 145–151 (1991)
30. Liu, H.X., Wu, X., Ma, W., Hu, H.: Real-time queue length estimation for congested signalized intersections. Trans. Res. Part C Emerg. Technol. **17**(4), 412–427 (2009)
31. Merseguer, J., Campos, J., Bernardi, S., Donatelli, S.: A compositional semantics for UML state machines aimed at performance evaluation. In: DES. pp. 295–302. IEEE (2002)
32. Nielsen, F.: On a generalization of the jensen-shannon divergence and the js-symmetrization of distances relying on abstract means. arXiv preprint arXiv:1904.04017 (2019)
33. Object Management Group: UML Profile for MARTE: Modeling and Analysis of Real-Time Embedded systems v1.0 (2009)
34. Object Management Group: OMG System Modeling Language (2017)
35. Paolieri, M., Biagi, M., Carnevali, L., Vicario, E.: The ORIS tool: quantitative evaluation of non-Markovian systems. IEEE Trans. Softw. Eng. **47**(6), 1211–1225 (2021)
36. Pooley, R., King, P.: The unified modelling language and performance engineering. IEE Proc.-Software **146**(1), 2–10 (1999)
37. Pyramis Library: https://github.com/oris-tool/pyramis (2025)
38. Radio Technical Commission for Aeronautics: DO-178B, Software Considerations in Airborne Systems and Equipment Certification (2012)
39. Reibman, A., Trivedi, K.: Numerical transient analysis of Markov models. Comput. Oper. Res. **15**(1), 19–36 (1988)
40. Runeson, P., Höst, M.: Guidelines for conducting and reporting case study research in software engineering. Empir. Softw. Eng. **14**(2), 131–164 (2009)
41. Russell, N., Ter Hofstede, A.H., Van Der Aalst, W.M., Mulyar, N.: Workflow control-flow patterns: A revised view (2006)
42. Schmidt, D.C.: Model-Driven Eng. Comput.-IEEE Comput.Soc. **39**(2), 25 (2006)

43. SIRIO Library: https://github.com/oris-tool/sirio (2024)
44. Tribastone, M., Gilmore, S.: Automatic translation of UML sequence diagrams into PEPA models. In: 2008 Fifth International Conference on Quantitative Evaluation of Systemss. pp. 205–214. IEEE (2008)
45. Trowitzsch, J., Zimmermann, A., Hommel, G.: Towards quantitative analysis of real-time UML using stochastic Petri nets. In: I19th IEEE International Parallel and Distributed Processing Symposium pp. 7–pp. IEEE (2005)
46. Tukey, J.W., et al.: Exploratory data analysis, vol. 2. Reading, MA (1977)
47. Vicario, E., Sassoli, L., Carnevali, L.: Using stochastic state classes in quantitative evaluation of dense-time reactive systems. IEEE Trans. Softw. Eng. $35(5)$, 703–719 (2009)
48. Vijaykumar, N.L., Carvalho, S., Andrade, V., Abdurahiman, V.: Introducing probabilities in statecharts to specify reactive systems for performance analysis. Comp. & Oper. Res. $33(8)$, 2369–2386 (2006)
49. Vogl, U., Siegle, M.: A new approach to predicting reliable project runtimes via probabilistic model checking. In: Reinecke, P., Di Marco, A. (eds) Computer Performance Engineering. European Workshop on Performance Engineering. pp. 117–132. Springer (2017). https://doi.org/10.1007/978-3-319-66583-2_8
50. Zheng, Z., Trivedi, K.S., Qiu, K., Xia, R.: Semi-Markov models of composite web services for their performance, reliability and bottlenecks. IEEE Trans. Ser. Comput. $10(3)$, 448–460 (2015)
51. Zheng, Z., Lyu, M.R.: Ws-dream: A distributed reliability assessment mechanism for web services. In 2008 IEEE International Conference on Dependable Systems and Networks With FTCS and DCC (DSN). pp. 392-397, IEEE (2008)

# CyberSecurity Performance Evaluation Through Attack Block Diagrams

Salvatore Distefano$^{(\boxtimes)}$ 

Mathematics and Computer Science Department, University of Messina,
Messina, Italy
`sdistefano@unime.it`
`https://unime.unifind.cineca.it/resource/person/6473`

**Abstract.** Security modeling is a critical aspect in protecting computing systems. Different techniques, such as attack trees and attack graphs, among others, have been developed to visualize and analyze potential threats and vulnerabilities. However, these techniques have limitations in capturing the complexity and dynamics of current computing systems, often distributed, networked, virtualized and software defined. To address these limitations, this paper introduces a novel security modeling technique, Attack Block Diagrams (ABD), aiming to provide a more flexible and comprehensive approach to security modeling and performance evaluation. ABD combine elements of attack trees, attack graphs, and reliability block diagrams (RBD) into a novel framework for investigating and assessing potential attack paths in complex computing systems. In this paper a formal definition of ABD and some examples taken from the literature are provided to show the advantages of the proposed formalism against other techniques such as attack trees and attack graphs.

**Keywords:** Quantitative Security · Cybersecurity Dynamics · Attack Modeling Techniques · Time-to-attack · Attack Trees and Graphs

## 1 Introduction

Cybersecurity science, engineering and technologies primarily focus on threat analysis, vulnerability management, security policy design, and incident response [14]. The advancement of this field critically depends on the specific definition, rigorous development, and effective application of cybersecurity metrics, models, and associated analysis techniques. In particular, this is required for investigating cybersecurity risks, understanding attack and defense processes, and characterizing their inherent dynamic behaviors. *Cybersecurity dynamics* investigates intricate processes influenced by various spatio-temporal dynamic phenomena, including sophisticated attack strategies, mutable system configurations, and human factors [27].

Attack modeling techniques (AMT) systematically represent and visualize the sequence or combination of events that can culminate in a successful cyber-


L. Carnevali and J. Doncel (Eds.): EPEW 2025, LNCS 15657, pp. 147–166, 2026.
https://doi.org/10.1007/978-3-032-16345-5_11

attack on a computing system or network. These techniques are broadly categorized into use case-based, temporally-focused, and graph-based methodologies [13]. Use case-based techniques represent attack scenarios from the perspective of an attacker goals and system interactions, often employing textual narratives or structured diagrams to describe step-by-step attack sequences and their preconditions and postconditions. Temporally-focused methods specifically address the time-dependent nature of attacks, considering factors such as time-to-compromise and the dynamics of event sequences. Among the most common and widely used AMT are attack trees and attack graphs. Both methodologies mathematically and graphically describe the workflow of events for a successful cyber-attack. Attack trees offer a hierarchical depiction of potential attacks against a system, enabling quantitative security evaluation and highlighting critical vulnerabilities [25]. Attack graphs are directed graphs that identify potential pathways for an attacker to violate a security policy, revealing causal relationships between events [21].

Beyond these mainstream approaches, prior research has explored alternative formalisms, including combinatorial models such as fault trees and block diagrams. In [5] a formal language for attack-defense trees using block diagrams is proposed, and several attempts to adopt fault trees for attack modeling are reported in [13,19]. In [23] the authors also combined a block diagram model with Markov and a multifragmented models to represent Cloud system availability affected by network attacks.

Inspired by these works and by reliability block diagrams (RBD) [24], this paper proposes Attack Block Diagrams (ABD) to further improve AMT. ABD represent system components as blocks and depict their dependencies and interactions through dynamic, adaptable connections. ABD identify potential attack paths as sequences of interconnected blocks. While previous works [19,23] share a similar visual metaphor, our proposed Attack Block Diagram (ABD) formalism differs significantly by grounding its methodology in System Reliability Theory [24]. This allows for a direct and systematic application of quantitative analysis techniques from a mature field, providing a structured framework for evaluating security metrics in a manner distinct from general graph-based approaches. The ABD "block" is the equivalent of a component in a Reliability Block Diagram (RBD), allowing for the rigorous analysis of attack paths as successful paths in the system security posture. Compared to attack trees, ABD thus enable greater flexibility in representing dynamic dependencies and the potential for more comprehensive quantitative analysis. Compared to attack graphs, ABD provide insights into the likelihood and impact of attack scenarios, in addition to identifying attack paths. ABD enhance modeling flexibility, improve risk assessment, and allow more intuitive visualization of attack surfaces.

This paper defines and formalizes the Attack Block Diagram (ABD) formalism and provides a comparative analysis with attack trees and attack graphs. The remainder of this paper is organized as follows: Sect. 2 introduces preliminary concepts, including cybersecurity basics, attack modeling techniques, and related work. Section 3 presents the definition and formalization of Attack Block

Diagrams. Section 4 provides some examples of ABD. Finally, Sect. 6 concludes the paper and outlines future work.

## 2    Preliminary Concepts

A computing system is vulnerable to cybersecurity risks, or threats. Threats are triggered by unauthorized activities, termed attacks. Exploitation results from actions compromising system security. Threat identification and attack analysis are therefore crucial for mitigating these risks. These investigations drive proactive and reactive cybersecurity techniques.

Graph-based attack modeling visualizes potential cyber threats. Attack trees and attack graphs are widely used. Attack trees (AT) detail attack paths in a hierarchical structure. They decompose a high-level attack goal into sub-goals through Boolean logical relationships among attack events expressed by gates. In AT attack events and gates are vertices of the tree/graph, while edges represent logical immediate transitions from an event to another one in the AT. Attack trees thus implement a top-down approach enforcing clarity towards a goal-oriented methodology. Attack trees are valuable for qualitative and quantitative analysis. However, their scope is limited, and they are often static. They can become unwieldy.

Attack graphs (AG) depict potential attack paths using directed acyclic graphs. They depict system state sequences and transitions from attacker exploits. Vertices thus represent the system state, while edges represent attack events. Thereby, AG identify critical paths compromising the system security. They enhance security by identifying vulnerabilities and trust relationships, expressed by Boolean logic among the edges. The AG approach starts from the bottom by investigating the initial state of the system components, then it explores potential attack events and paths leading to the system exploitation, therefore, it is bottom-up. However, AG can be complex, leading to state-space explosion, thus suffering from scalability issues and high computational costs [26].

Cybersecurity attack metrics and key performance indicators quantify and assess the impact and likelihood of cyberattacks. These metrics provide a basis for prioritizing security measures and evaluating the effectiveness of attack models and defense countermeasures. The main metric usually adopted in attack modeling is the attack probability $p$, i.e., the likelihood of a successful attack. However, cybersecurity dynamics [27] highlights the evolving nature of threats, vulnerabilities, and defenses in cyberspace. Attack models must adapt to these dynamics. This includes the changing attack patterns, the discovery of new vulnerabilities, and the development of new security measures. Effective attack modeling should account for the time-dependent nature of attacks and the potential for attackers to change their strategies in response to defenses.

To this end, a specific elementary attack event $i$ is stochastically characterized by the *time-to-attack/compromise* random variable $x_i$ first introduced in [18] and then refined by [15,17], where $F_i(t)$ of Eqs. (1) and (2)

$$F_i(t) : \mathbb{R} \to [0,1] \subset \mathbb{R} \tag{1}$$

is the cumulative distribution function (cdf) of $x_i$, i.e.

$$F_i(t) \; = \; Pr\{\text{the } i \text{ attack event occurs in } [0,t]\} \; = \; Pr\{x_i \leq t\}. \tag{2}$$

At time $t = 0$, the system is assumed to be not under attack, i.e. $F_i(0) = 0$. An attack, if any, cannot start before $t = 0$, but later at $t > 0$. The mean time to attack/compromise (MTTC) is a critical security metric representing the expected time until the component, block or system is compromised or the attack succeed. Given the time to compromise cdf for block $i$ $F_i(t)$, the MTTC is defined as the integral of the survivability function $S(t) = 1 - F(t)$, from $t = 0$ to infinity as defined by Eq. (3)

$$MTTC = \int_0^\infty S(t)dt. \tag{3}$$

A feasible solution to include stochastic characterizations of attack dynamics into specific attack modeling techniques can be inspired by similar contexts, such as the reliability engineering one. In reliability engineering, combinatorial models, primarily Fault Trees (FT) and Reliability Block Diagrams (RBD), are often exploited to represent system reliability [24]. There are even some successful attempts to adapt FT to model attack processes [13], even combined into attack fault trees (AFT) [11]. But despite the RBD flexibility, to the best of our knowledge, there are very few attempts to adapt the RBD formalism to attack process modeling [8]. This can be, however, a good solution to deal with almost all the open issues and challenges of AT and AG, since RBD is a quite intuitive formal notation based on a solid mathematical framework, which also covers system dynamics. RBD, indeed, is a well-established modeling technique in reliability engineering used to analyze system reliability based on the structural arrangement and failure characteristics of its components [24]. RBD represent system components as blocks connected in series, parallel, or more complex configurations, where the functioning of the system depends on the functioning of these blocks according to the defined connections. This formalism allows for the quantitative assessment of system reliability and availability. The principles of representing system dependencies and enabling quantitative analysis inherent in RBD provide a source of inspiration for developing new security modeling techniques.

In such a direction, the inclusion of time as a critical metric has been a recurrent theme in the literature [2]. Several attempts have been made to estimate the time-to-attack/compromise within the AT and AG frameworks [16], even exploiting different models such as PEPA [1] and non-homogenous continuous-time Markov models [3]. While valuable, these models often inherit the fundamental limitations of their underlying frameworks, such as scalability issues and limits in capturing dynamic changes. Similarly, moving target defense (MTD) approaches [4] model the dynamic evolution of defenses to increase attacker effort, even by exploiting AMT [7,10] or ML models [28]. This work complements these efforts

**Table 1.** Comparison of AT, AG, and ABD

| Feature | Attack Trees | Attack Graphs | Attack Block Diagrams |
|---|---|---|---|
| *Structure* | Directed graph (Acyclic Hierarchical Tree) | Directed Graph | Directed Graph |
| *Vertex* | Attack Event and Gate | System state | Block (representing attack components or events) |
| *Edge* | Flow Connection | Attack Event | Block Structure |
| *Logic* | Boolean | Boolean Edge Relationships | Structure |
| *Dependency* | S-independent Events | S-independent Events | S-independent Blocks |
| *Approach* | Top-down | Bottom-up | Hybrid (both top-down and bottom-up or mixed) |
| *Analysis* | Logic, Probabilistic | Logic, Probabilistic | Logic, Probabilistic, Stochastic |
| *Strengths* | Clarity, goal-oriented, probabilistic analysis [9, 22] | Comprehensive coverage, integration [12] | Clarity, stochastic analysis, modularity, hybrid approach |
| *Weaknesses* | Limited scope, static representation [6, 22] | Complexity, computational cost [26] | Novelty, lack of specialized tools |
| *Best Applications* | Simple systems, specific goals | Complex systems, all possible paths | Systems with complex structure, risk assessment |
| *Standardization* | No | No | No |

by introducing a formalism that inherently accounts for cybersecurity dynamics [27] by allowing the connections between blocks to adapt based on the system states or attacker actions. This provides a robust foundation for analyzing how the attack surface evolves, a core challenge in modern computing systems.

Table 1 presents a structured comparison of Attack Trees (AT), Attack Graphs (AG), and Attack Block Diagrams (ABD). All three methodologies are graph-based, representing attacks as interconnected elements. However, Attack Trees (AT) are distinguished by their acyclic, hierarchical structure, resembling a tree with branches that do not loop back on themselves. This hierarchical organization makes AT suitable for illustrating how a high-level attack goal can be broken down into progressively smaller, more manageable sub-goals, but may be a limit for their applications to iterative, incremental attack processes and workflows. The fundamental building blocks also differ. AT use attack events (specific actions taken by an attacker) and gates (logical operators like AND or OR, showing how sub-goals combine to achieve a larger goal). AG, on the other hand, revert somehow the AT logic, associating system states with vertices, representing the condition of the system at a particular point during an attack, while the state change transition event is represented by edges. ABD employ blocks, which can represent either attack components (parts of the system under attack) or attack events. In AT, flow connections show how attack events and gates are related.

# 3    Attack Block Diagrams

The Attack Block Diagram (ABD) method is a practical security modeling tool for real and abstract computing systems. ABD provides a systemic perspective on the cybersecurity attack process. It graphically represents system architecture security, depicting relationships between physical and logical subsystems, components, or events that lead to potential attack paths. ABD intuitive readability makes it understandable to various stakeholders, including both technical and non-technical personnel. Security engineers can efficiently construct, validate, and adapt ABD, facilitating cross-domain communication. The ABD logic diagram reveals how component attack events can compromise a system, or conversely, how secure components maintain system integrity. ABD are based on system design, procedures, controls, and attack consequence analysis. An ABD block represents a component or an attack event. A successful attack event on a component activates or modifies its block or connections. The system is compromised if enough attack events activate an attack path between ABD terminal points. If no attack paths are active, the system has a degree of security against the modeled attacks.

## 3.1    Blocks

The basic element, the unit of an attack block diagram is a block. It represents an *"attack"* event which may directly or indirectly contribute to a successful attack or exploitation of the system. In an ABD, attack blocks, components, or events are *binary*, i.e., they are characterized by two states: success/compromised or failure/not compromised. Thus, a generic block $b_i$ is associated with and represents a specific elementary attack event, stochastically characterized by the time-to-attack or time-to-compromise random variable $x_i$ and its cdf $F_i(t) : \mathbb{R} \to [0, 1] \subset \mathbb{R}$ as defined above. This is an assumption inherited by ABD from AT and AG, where an attack event is binary, can only succeed or fail. However, 3 or more states, for example inspired by the cyber kill chain or similar models on attack process and dynamics, can absolutely make sense and thus become ideas for future work. The time-to-attack cdf $F_i(t)$ is the main performance metric here defined to stochastically characterize an attack block, system and/or process.

## 3.2    System

An ABD is a connected graph aiming to identify all the paths of basic attack events leading to successful system attacks. An ABD consists of the following components: i) the blocks representing system components, elements, or events that may lead to successful system attacks, as specified above; and ii) connections modeling the interactions between blocks in the systemic view leading to successful system attacks. The main basic assumptions for an ABD are:

1. The system, as well as each block, is binary, characterized by only two states, that is, a successful or a failed attack.
2. The ABD represents a successful system attack by an attack path, i.e., a sequence of successful basic attack events or blocks.
3. ABD block attack events are *statistically independent*, i.e. the probability of a successful attack on block $b_i$ in the $[0, t)$ time interval is $F_i(t)$, not related to the probability of a successful attack on block $b_j$ in the same time frame $F_j(t)$ for all distinct system blocks $b_i$ and $b_j$ among the $n$ blocks. Formally:

$$P(\text{Attack on } b_i \text{ in } [0, t) | \text{Attack on } b_j \text{ in } [0, t)) =$$

$$= P(\text{Attack on } b_i \text{ in } [0, t)) = F_i(t) \quad \forall i \neq j \in [1, .., n]$$

This assumption implies that the occurrence of a successful attack on one component does not influence the likelihood of a successful attack on any other distinct component within the system model represented by the ABD.

4. The system attack is stochastically represented by a cdf $F(t)$ depending on all component-block attack cdf

$$F(t) : \mathbb{R} \to [0, 1] \subset \mathbb{R} = G(F_1(t), ..., F_n(t))$$

The main goal of an ABD is to identify all the attack paths leading to a successful system attack, and to assess the attack probability at time $t$. Attack paths represent sequences of blocks connected by lines characterizing system attack scenarios. ABDs can be visualized as directed graphs, where nodes represent blocks and edges represent connections.

Formally, an ABD is a connected directed graph $G(\mathbf{B}, \mathbf{E})$, composed of a set of vertices or blocks $b_i \in \mathbf{B}$ and a set of edges $e_k \in \mathbf{E}$ that represent relationships between the blocks. The overall graph structure can be thus expressed as a pair of the form $G(\mathbf{B}, \mathbf{E})$ where:

- $\mathbf{B}$ is a finite nonempty set of blocks. It contains at least 2 *terminal* states $s, e \in \mathbf{B}_0 \subseteq \mathbf{B}$, representing the *start* $s$ and *end* $e$ terminal blocks, thus $|\mathbf{B}| = n \geq 2$. Terminal blocks are fictitious blocks characterized by always successful attack cdf $F_s(t) = F_e(t) = 1 \, \forall t \in \mathbb{R}^+$.
- $\mathbf{E} \subseteq \mathbf{B} \times \mathbf{B}$ is an edge transition relation, represented as an ordered pair of blocks $(b_i, b_j) \in \mathbf{E}$ where the order models the edge direction (outgoing) from $b_i$ (incoming) to $b_j$ since the ABD is a directed graph, and thus $(b_i, b_j) \in \mathbf{E} \nRightarrow (b_j, b_i) \in \mathbf{E}$.

As a directed graph, an ABD is navigated from the start terminal $s$ to the end one $e$. An ABD *attack path* $p \in \mathbf{P}$ of *length* $1 \leq k \leq n$ from a block $b_1 \in \mathbf{B}$ to a block $b_k \in \mathbf{B}$ is a sequence of blocks $(b_1, \ldots, b_i, \ldots, b_k)$ such that:

1. $b_i \in \mathbf{B} \; \forall \, i \in \{1, \ldots, k\}$.
2. $(b_i, b_{i+1}) \in \mathbf{E} \; \forall \, i \in \{1, \ldots, k-1\}$.

The blocks $b_1$ and $b_k$ are the *endpoints* of the path, and $b_1$ is the initial block and $b_k$ is the final one, while $b_2, \ldots, b_{k-1}$ are internal blocks. In an ABD, start $s$ and end $e$ blocks are reachable, i.e. there exists at least a path $p \in \mathbf{P}$ between them. Furthermore, for any vertex or block $b_i \in \mathbf{B} - \mathbf{B_0}$, there must exist a path from the start block $s$ to $b_i$, i.e. $(s, \ldots, b_i) \in \mathbf{P}$, and also a path from $b_i$ to the end block $e$, i.e. $(b_i, \ldots, e) \in \mathbf{P}$. Conversely, start and end blocks are reachable by any other node. This also implies that there exists at least a path $p_{b_i}$ from $s$ to $e$ through $b_i$, formally:

$$\forall b_i \in \mathbf{B} - \mathbf{B_0} \rightarrow \exists p_{b_i} \in \mathbf{P} | p_{b_i} = (s, \ldots, b_i, \ldots, e).$$

## 3.3   Structured ABD

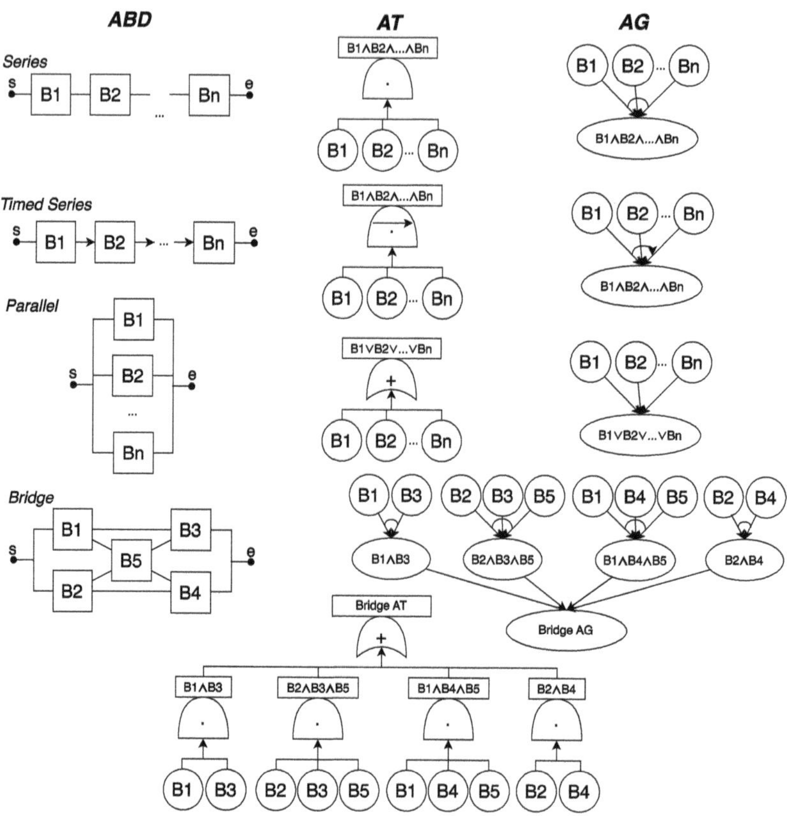

**Fig. 1.** Comparison of ABD structures with Attack Trees (AT) and Attack Graphs (AG) corresponding notation elements.

ABD is a quite powerful formalism, able to represent a wide variety, potentially any combination of statistically independent events leading to a successful system attack in multiple ways. To restrict this degree of freedom that may also lead to ambiguity and/or semantically unsafe or not well-formed model, structured ABD are specified from well-known and well-formed structures, as done for AT and AG by specific gates or connections, respectively. Analogously to RBD, basic structured ABD rely on series, parallel, and combined, even nested, series-parallel structures as shown in Fig. 1, taken from reliability engineering [24]. Furthermore, it is important to highlight that structured ABD can also include replicated blocks, a common practice for AT and AG as shown at the bottom of Fig. 1.

In structured ABD, system components or security controls are represented as blocks, and their arrangement illustrates how successful attacks on these elements can lead to system compromise. Similarly to RBD, ABD exploit specific structural arrangements to model the logical dependencies of security as follows.

*Series Structure.* An ABD block with a series structure is composed of $n$ components arranged sequentially. For an attack on the series block to succeed (i.e., the series is in a successful attack state), all $n$ component blocks within the series must be in a successful attack state, meaning that an attack on each individual block must succeed. If even only one block in the series is not successfully attacked, the entire series block is considered to be safe or not successfully attacked. The block only enters a successful attack state (attack succeeds) if all $n$ components are successfully attacked. This structure models scenarios where a chain of attacks must all be successful for the system to be in an unsafe-successful attack state, a redundant configuration, which is equivalent to an AT AND gate or the AG AND connection as shown in the top of Fig. 1. A variant of the series structure is the *timed series*, which is equivalent to a Priority-And gate or event that successfully triggers an attack event if and only if all the basic events occur in the order specified by the gate (left to right). This is a widely adopted gate/event in attack modeling.

*Parallel Structure.* An ABD block with a parallel structure consists of $n$ blocks arranged in parallel. The attack on a parallel block is successful (the parallel is in a successful attack state) if at least one out of the $n$ components is successfully attacked. The parallel block thus enters a successful attack state (attack succeeds) if only one component is successfully attacked. This structure represents weak security mechanisms, where a single attack to a block or part of the system leads to the system successful attack, equivalent to an AT OR gate or the AG OR connection as shown in the middle of Fig. 1.

*Bridge Structure.* An ABD block with a bridge structure is a specific arrangement of 5 blocks $b_1, b_2, b_3, b_4, b_5$ as illustrated on the bottom of Fig. 1. The bridge structure is considered to be in a successful attack state if at least one of the four following conditions is satisfied:

1. Blocks $b_1$ and $b_3$ are in a successful attack state.
2. Blocks $b_1$, $b_4$ and $b_5$ are in a successful attack state.
3. Blocks $b_2$, $b_3$ and $b_5$ are in a successful attack state.
4. Blocks $b_2$ and $b_4$ are in a successful attack state.

This can model more intricate network or system topologies with alternative secure pathways. The equivalent AT and AG models are shown in Fig. 1, highlighting the higher compactness of ABD than AT and AG due to the systemic view. A practical cybersecurity attack example is a zero-day exploit in a firewall or an intrusion prevention system (IPS). Imagine an attacker goal is to gain access to a critical file server inside a corporate network in two ways: i) However, a bridge component (B5) represents a single, highly effective attack that can bypass these two distinct paths, e.g. a zero-day exploit in the company central firewall. A successful execution of B5 would allow the attacker to move directly from the compromised employee machine (after B1) to the internal network, bypassing the need for a privilege escalation (B3) unlocking the web server exploitation internal network (B4). The bridge B5 connects the initial block B1 directly to the network, effectively skipping the need to complete the full series of attacks both paths. i) Path A - The attacker initiates a social engineering attack (B1) to obtain employee credentials, and with such credentials, they can bypass the perimeter firewall and gain initial access to the network (B3); or Path B - The attacker could also try to exploit a vulnerability in a public-facing web server (B2), and from there, pivot to another server by exploiting an insecure configuration on a database server (B4). Normally, these two paths are completely independent. However, a bridge component, B5, represents a situation where a single, successful action can connect these separate avenues, e.g. the discovery and successful exploitation of a zero-day vulnerability in the Active Directory authentication service. A successful execution of B5 would allow the attacker, after completing the first step of Path A (B1, social engineering), to directly execute the second step of Path B (B4, exploiting insecure configuration on the database server) and viceversa B2 to B3. The bridge component B5 connects the endpoint of blocks B1/B2 directly to the starting point of blocks B4/B3, effectively bridging the two attack paths.

*Combined Series-Parallel Structure:* ABD can also represent more complex systems by combining series and parallel structures. This allows for the modeling of scenarios where some less critical, redundant, parts of the attack path require all blocks to be successfully attacked to trigger the successful attack of such parts (series), while others parts trigger an attack if only a block is successfully attacked (parallel). Analyzing such structures involves decomposing them into their constituent series and parallel elements, which can even be hierarchically nested.

### 3.4  ABD Structured Block Analysis

One of the key advantages of ABDs is their ability to provide a quantitative analysis of vulnerabilities and risks. By leveraging techniques from reliability engineering, ABDs can assess the likelihood of different attack scenarios and their potential impact on the system. This quantitative analysis can be used to prioritize security efforts and allocate resources more effectively. Specifically, to quantitatively assess the likelihood of a successful attack on an ABD block, described by the system attack cdf $F(t)$, the $G(\cdot)$ expression of $F(t) = G(F_1(t), ..., F_n(t))$ has to be specified given its block cdf $F_1(t), ..., F_n(t)$, assuming all blocks are statistically independent. In the following, it is specified for the ABD structures above described.

*Series Structure.* For a series block with $n$ independent components, the successful attack cdf on a series $(F_S(t))$ is specified by Eq. (4)

$$F_S(t) = \prod_{i=1}^{n} F_i(t). \tag{4}$$

A timed series requires a set of independent input events $B_1, B_2, \ldots, B_n$ to occur in a specific, predetermined sequence, one after another in sequence of events where the occurrence of one triggers the next to start. It is the equivalent of a sequential AND in attack and fault trees.

The ABD timed series block requires the sequence $B_1 \rightarrow B_2 \rightarrow \cdots \rightarrow B_n$, where the $(i)$-th event begins after the $(i-1)$-th event completion. The total time $T_{TS}$ is the sum of the independent event durations $(X_i)$.

$$T_{TS} = X_1 + X_2 + \cdots + X_n$$

The PDF of the total time $T_{TS}$ is the convolution of the individual pdf $(f_i)$:

$$f_{TS}(t) = f_1 * f_2 * \cdots * f_n(t)$$

The $F_{TS}$ CDF is the integral of the convolution of Eq. (5).

$$F_{TS}(t) = \int_0^t (f_1 * f_2 * \cdots * f_n)(x)\, dx. \tag{5}$$

If $B_i \sim \text{Exp}(\lambda_i)$ with distinct rates $\lambda_i$, the closed-form CDF is:

$$F_{TS}(t) = 1 - \sum_{i=1}^{n} \left( e^{-\lambda_i t} \prod_{j=1, j\neq i}^{n} \frac{\lambda_j}{\lambda_j - \lambda_i} \right).$$

*Parallel Structure.* An ABD parallel block with $n$ independent components is not compromised only if ALL $n$ components are not compromised or it is compromised when at least one component is compromised. The ABD parallel block probability that an attack is not successful $1 - F_P(t)$ can be therefore expressed by Eq. (6).

$$1 - F_P(t) = \prod_{i=1}^{n}(1 - F_i(t)) \quad \rightarrow \quad F_P(t) = 1 - \prod_{i=1}^{n}(1 - F_i(t)). \tag{6}$$

*Bridge Structure.* Assuming independence, this can be expanded using probability rules (e.g., inclusion-exclusion): Assuming independence, the probability of a successful attack $P_B$ for the 5-component bridge structure shown in the bottom of Fig. 1 can be expressed as:

$$P_B = P((b_1 \cap b_3) \cup (b_2 \cap b_4) \cup (b_1 \cap b_5 \cap b_4) \cup (b_2 \cap b_5 \cap b_3))$$

Expanding this using probability rules (e.g., inclusion-exclusion) the bridge structure cdf $F_B(t)$ can be expressed by Eq. (7).

$$\begin{aligned}
F_B(t) = {}& F_1(t)F_3(t) + F_2(t)F_4(t) + F_1(t)F_5(t)F_4(t) + F_2(t)F_5(t)F_3(t)+ \\
& - F_1(t)F_3(t)F_2(t)F_4(t) - F_1(t)F_3(t)F_1(t)F_5(t)F_4(t)+ \\
& - F_1(t)F_3(t)F_2(t)F_5(t)F_3(t) - F_2(t)F_4(t)F_1(t)F_5(t)F_4(t)+ \\
& - F_2(t)F_4(t)F_2(t)F_5(t)F_3(t) - F_1(t)F_5(t)F_4(t)F_2(t)F_5(t)F_3(t)+ \\
& + \text{higher order intersection terms.}
\end{aligned} \tag{7}$$

A more systematic approach using minimal cut sets or path tracing is generally recommended in the analysis.

*Combined Series-Parallel Structure* analysis involves iteratively applying the formulas for series and parallel blocks to the decomposed sub-structures.

## 4    Examples

### 4.1    SSHD Buffer Overflow

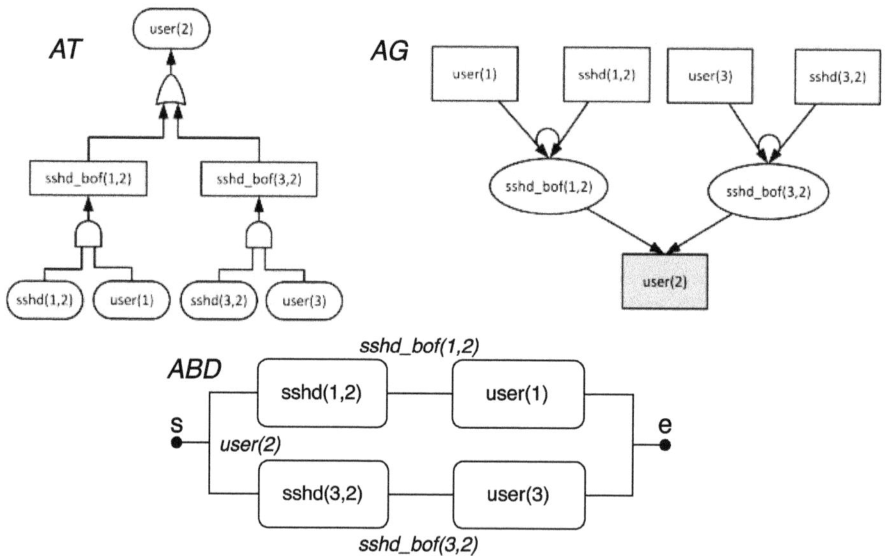


**Fig. 2.** SSHD Buffer Overflow Example AT, AG and ABD from [13].

The example of Fig. 2, taken from [13], describes and compares two Attack Modeling Techniques (AMT): Attack Trees (AT) and Attack Graphs (AG), against the proposed Attack Block Diagram (ABD) for a cyber-attack scenario. This example outlines a sequence of exploit executions, specifically a buffer overflow vulnerability in the Secure Shell Daemon (*sshd_bof*), targeting a series of networked host computing devices. The successful exploitation of this vulnerability on each host culminates in the acquisition of user-level privileges (*user*) on those systems. Furthermore, the example highlights a critical prerequisite for a successful exploit: the presence and operational status of the Secure Shell Daemon (sshd) on the target host. Graphically, the ABD shows better than the other formalisms the attack paths and the dynamics of cyber-attacks, thereby enhancing situational awareness and facilitating a deeper understanding of attack vectors and dependencies. It also allows obtaining the successful attack probability cdf $F_{user(2)}(t)$ given the (s-independent) component ones as follows:

$$F_{user(2)}(t) = 1 - (1 - F_{sshd(1,2)}(t)F_{user(1)}(t))(1 - F_{sshd(3,2)}(t)F_{user(3)}(t))$$

## 4.2   Identity Theft

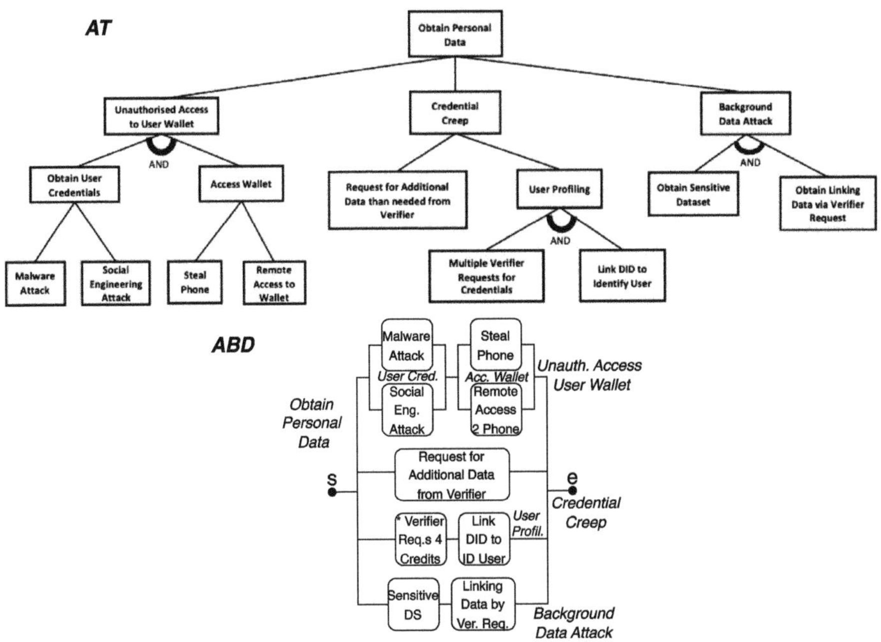


**Fig. 3.** Identity Theft example AT and ABD from [20].

This example, taken from [20] focused on identity thefts. Figure 3 depicts the AT and ABD models of potential attack vectors for identity theft, wherein a malicious actor aims to acquire personal data by exploiting identified vulnerabilities within the Self-Sovereign Identity (SSI) architecture assets [20]. Each event requires a rigorous evaluation of its probability of success and associated severity, to determine the corresponding attack cdf. The depicted scenarios encompass unauthorized access to personal data by an attacker or other stakeholders, or access without explicit user consent. Specifically, vulnerabilities inherent in infrastructure assets, such as authentication deficiencies in network hosts, can enable threat actors with direct access to a personal wallet to exfiltrate stored personal data or credentials. The inherent architecture and functionality of the SSI system constitute a potential attack surface for compromising personal data and credentials. Furthermore, the credential verification mechanism is sensitive to exploitation via the credential creep technique, facilitating the collation of personal data through excessive information requests. This attack vector involves a verifier soliciting data beyond the minimal requirements for claim verification. Repetitive verifier requests can similarly be used to aggregate personal data and establish linkages to the targeted user Decentralized Identifier.

**Table 2.** Cumulative Distribution Functions (cdf) of Attack Events

| Cdf | Event | Formula |
|---|---|---|
| $F_{MA}(t)$ | Malware Attack | Given as basic event |
| $F_{SEA}(t)$ | Social Engineering Attack | Given as basic event |
| $F_{SP}(t)$ | Steal Phone | Given as basic event |
| $F_{RA}(t)$ | Remote Access to Wallet | Given as basic event |
| $F_{RAD}(t)$ | Request for Additional Data | Given as basic event |
| $F_{MR}(t)$ | Multiple Verifier Requests for Credentials | Given as basic event |
| $F_{LDU}(t)$ | Link DID to Identify User | Given as basic event |
| $F_{OSD}(t)$ | Obtain Sensitive Dataset | Given as basic event |
| $F_{OLD}(t)$ | Obtain Linking Data via Verifier Request | Given as basic event |
| $F_{OUC}(t)$ | Obtaining User Credentials | $1 - (1 - F_{MA}(t))(1 - F_{SEA}(t))$ |
| $F_{AW}(t)$ | Access Wallet | $1 - (1 - F_{SP}(t))(1 - F_{RA}(t))$ |
| $F_{UA}(t)$ | Unauthorized Access to User Wallet | $F_{OUC}(t)F_{AW}(t)$ |
| $F_{UP}(t)$ | User Profiling | $F_{MR}(t)F_{LDU}(t)$ |
| $F_{CC}(t)$ | Credential Creep | $1 - (1 - F_{RAD}(t))(1 - F_{UP}(t))$ |
| $F_{BDA}(t)$ | Background Data Attack | $F_{OSD}(t)F_{OLD}(t)$ |

A comprehensive cybersecurity risk assessment requires the evaluation of all plausible attack paths identified by the AT and the ABD to derive informed conclusions regarding the overall risk profile associated with identity theft within the SSI ecosystem and to inform the implementation of appropriate mitigation strategies. Assuming to know all the basic events attack cdf as reported in Table 2, it is possible to get the overall system time-to-attack cdf. The root goal

Obtain Personal Data cdf $F_{OPD}$ is given by the cdf of the three main attack vector parallel and thus expressed as a function of basic attack event cdfs as follows:

$$F_{OPD}(t) = 1 - (1 - F_{UA}(t))(1 - F_{CC}(t))(1 - F_{BDA}(t))$$

$$= 1 - (1 - F_{OUC}(t)F_{AW}(t))(1 - F_{RAD}(t))(1 - F_{UP}(t))(1 - F_{OSD}(t)F_{OLD}(t))$$

$$= 1 - [1 - (1 - (1 - F_{MA}(t))(1 - F_{SEA}(t)))(1 - (1 - F_{SP}(t))(1 - F_{RA}(t)))] \cdot$$

$$\cdot (1 - F_{RAD}(t))(1 - F_{MR}(t)F_{LDU}(t))(1 - F_{OSD}(t)F_{OLD}(t)).$$

## 5  Safety vs. Security Antagonism

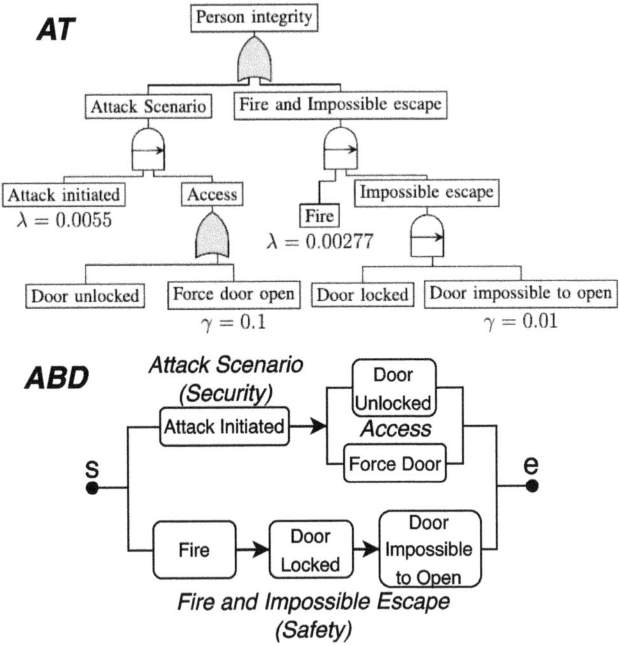


**Fig. 4.** Safety-Security example AT and ABD from [11].

This example, taken from literature [11] and shown in Fig. 4, represents a classic conflict between safety and security in the design of an emergency exit door. The core design choice revolves around whether the door should remain always locked or always unlocked. From a safety perspective, it is critical that the door remains unlocked. However, from the security perspective, an open door is a security vulnerability that an attacker could exploit. The top event, Person Integrity, can be reached through two primary, competing paths (modeled as the final aggregation block): The Safety Failure Path - this path succeeds if there is a fire and the door remains locked. The Security Breach Path - this path succeeds if an attack is initiated and access is obtained via breaking the door. The model captures the initial status of the door using two leaves—Door Locked and Door Unlocked—with the restriction that the locked door status $(1 - p)$ forces the attacker down a time-dependent path (such as Force Open the Door), whereas the unlocked door status $(p)$ leads to immediate success (Walk-in). The following Table 3 details the event rates $(\lambda)$ for the exponentially distributed events and the probability $(p)$ associated with the initial door state.

**Table 3.** ABD Model Parameters: Rates $(\lambda)$ and Probability $(p)$

| Event Name | Symbol | Type | Value (hour$^{-1}$) |
|---|---|---|---|
| Fire | $\lambda_F$ | Rate $(\lambda)$ | 0.00277 |
| Door Impossible to Open | $\lambda_{DI}$ | Rate $(\lambda)$ | 0.01 |
| Force Door Open | $\lambda_{FD}$ | Rate $(\lambda)$ | 0.1 |
| Attack Initiated | $\lambda_{AI}$ | Rate $(\lambda)$ | 0.0055 |
| Door Unlocked State (Design Choice) | $p$ | Probability | 0.5 |

From a stochastic viewpoint, the door unlocked event is characterized by $F_{DU} = p$ and locked one by $F_{DL} = 1 - F_{DU} = 1 - p$. The access parallel block is thus stochastically characterized by

$$F_A = 1 - (1 - F_{DU})(1 - F_{FD}) = 1 - (1 - p)(1 - F_{FD}) = p + (1 - p)F_{FD},$$

while the timed series attack scenario block cdf $F_{AS}$ is given by the attack initiated and the access event timed series, thus resulting in the convolution of their cdf:

$$F_{AS}(t) = F_{AI} * F_A = 1 - \frac{\lambda_{FD} - p\lambda_{AI}}{\lambda_{FD} - \lambda_{AI}}e^{-\lambda_{AI}t} + \frac{(1 - p)\lambda_{AI}}{\lambda_{FD} - \lambda_{AI}}e^{-\lambda_{FD}t}.$$

In the other path, the impossible escape event cdf is $F_{IE} = (1-p)F_{DI}$, while the full timed series block cdf $F_{FIE}$ can be expressed as

$$F_{FIE} = F_F * F_{IE} = (1 - p)(1 - \frac{\lambda_{DI}}{\lambda_{DI} - \lambda_F}e^{-\lambda_F t} + \frac{\lambda_F}{\lambda_{DI} - \lambda_F}e^{-\lambda_{DI}t}).$$

Thereby, the overall person integrity parallel block can be stochastically characterized by the $F_{PI}$ cdf as follows:

$$F_{PI} = 1 - (1 - F_{AS})(1 - F_{FIE}) = 1 - (1 - (1 - \frac{\lambda_{FD} - p\lambda_{AI}}{\lambda_{FD} - \lambda_{AI}}e^{-\lambda_{AI}t} + \frac{(1 - p)\lambda_{AI}}{\lambda_{FD} - \lambda_{AI}}e^{-\lambda_{FD}t})).$$

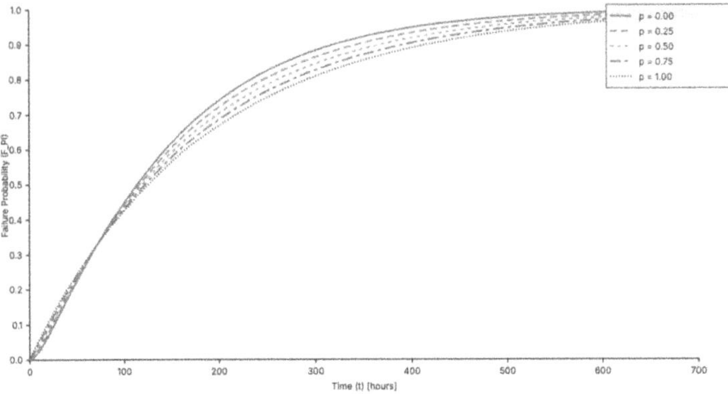

**Fig. 5.** Safety-Security example time-to-compromise cdf plots on different $p$ values, $\lambda_{FD} = 0.1$ and $\lambda_{DI} = 0.01$.

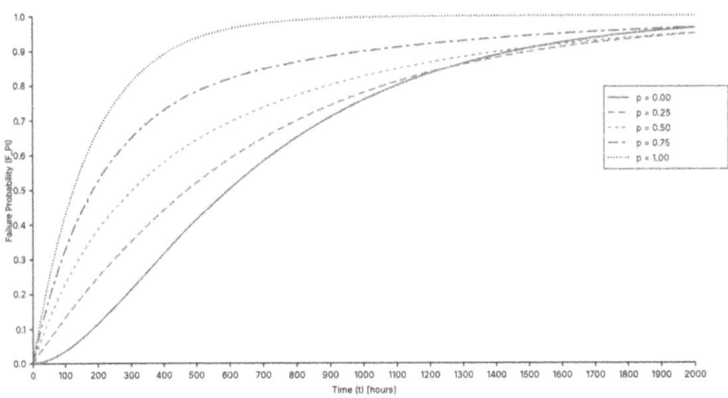

**Fig. 6.** Safety-Security example time-to-compromise cdf plots on different $p$ values, $\lambda_{FD} = 0.001$ and $\lambda_{DI} = 0.001$.

$$\cdot (1 - (1-p)(1 - \frac{\lambda_{DI}}{\lambda_{DI} - \lambda_F}e^{-\lambda_F t} + \frac{\lambda_F}{\lambda_{DI} - \lambda_F}e^{-\lambda_{DI} t}))$$

The graphs shown in Fig. 5 plot the $F_{PI}$ trend in time. They investigate the trade-off between security and safety by showing $F_{PI}(t)$ across five different design configurations for the access door, controlled by the parameter $p$, assuming the design choice of the door state $(p)$ is the most significant factor driving the system risk profile. From a security viewpoint, the lowest risk configuration is, counter-intuitively, $p = 1$ characterizing the door locked which is also the safest configuration. This is due to the fact that, even if the door is unlocked, requiring the attacker to first initiate the attack and then use time to force the door open, the dynamic of such an attack $\lambda_{FD} = 0.1$, as well as the $\lambda_{DI} = 0.01$ one, are 1–2 order of magnitude greater than the other rates (fire and attack initiated) as shown in Table 3.

**Table 4.** MTTC for Different Initial Probabilities $p$

| Initial Probability ($p$) | MTTC (Hours) $\lambda_{FD} = 0.1,\ \lambda_{DI} = 0.01$ | MTTC (Hours) $\lambda_{FD} = \lambda_{DI} = 0.001$ |
|---|---|---|
| 0.00 | 150.24 | 734.89 |
| 0.25 | 158.57 | 678.66 |
| 0.50 | 166.60 | 567.74 |
| 0.75 | 174.35 | 402.13 |
| 1.00 | 181.82 | 181.82 |

The security trend in $p$ is confirmed by the MTTC values reported in the middle column of Table 4, where the door locked configuration ($p = 0$) has an MTTC (150.24 h) lower than the unlocked one ($p = 1$, 181.82 h). Reducing the model stiffness, i.e., by considering $\lambda_{FD} = 0.001$ and $\lambda_{DI} = 0.001$, the results shown in Fig. 6 and reported in the rightmost column of Table 4 are obtained. This demonstrates the effectiveness of keeping a *robust* door locked in terms of security, since the dynamic of the $p = 0$ $F_{PI}$ curve is almost 4 times, in average, slower than the $p = 1$ one (734.89 h vs 181.82 h), and 3 times slower than the $p = 0.5$ one (567.74 h vs 181.82 h), also showing a non linear behaviour in $p$. This analysis strongly suggests that maintaining a locked state (low $p$) is crucial for maximizing system security by leveraging the protective delay provided by a *good and robust enough* door physical barrier, if the effort and time to force it is high ($\lambda FD \sim 0.001$), otherwise ($\lambda FD \sim 0.1$) it is better to keep the door unlocked also for security purposes. In the former case, any decrease in $p$ rapidly shifts the system toward the less secure scenario but safer one. A trade-off between security and safety can be enforced based on the requirements by trading off the values of $0 < p < 1$ considering security-safety requirements.

## 6    Conclusions and Future Work

This paper proposed a novel formalism to represent cybersecurity attack processes, the Attack Block Diagrams (ABD). ABD introduces several advantages over traditional security modeling paradigms and attack modeling techniques. Specifically, ABD facilitate the representation of intricate attack paths that may not be readily or adequately captured by conventional methodologies such as attack trees or attack graphs. Furthermore, ABD enable the quantitative analysis of vulnerabilities and associated risks, thereby providing a basis for informed decision-making processes. The visual representation provided by ABDs also offers a clear and intuitive depiction of potential attack vectors, as shown by the examples taken from the literature.

Despite such advantages, ABD represent a relatively new technique, and several challenges require further investigation. Specifically, the representation of dynamic aspects and behaviors, relaxing the statistical-independence among attack blocks and related events to include dynamic-dependent behaviors is subject of ongoing and future work, as well as considering non-binary (3 or more states) blocks. The development of robust software tools to support the design, analysis, and visualization of ABD is crucial for their practical application. Furthermore, an important area of future research lies in exploring the integration of ABD with established security modeling techniques, such as threat modeling, cyber kill chain, and risk assessment frameworks.

**Acknowledgments.** This work is partially funded by European Union - Next generation EU - PNRR - Missione 4, Componente 2, Investimento 1.1 - Bando PRIN 2022 PNRR - Decreto Direttoriale n. 1409 del 14-09-2022 - Progetto RESILIENT, CUP J53D23015040001, project id. P2022S4TTP, and Next generation EU - PNRR, SERICS âĂŞ "SECURITY AND RIGHTS IN THE CYBERSPACE" project 3D-SEECSDE, CUP J33C22002810001, project id. PE00000014.

# References


1. Abraham, S.M.: Estimating mean time to compromise using non-homogenous continuous-time Markov models. In: 2016 IEEE 40th Annual Computer Software and Applications Conference (COMPSAC), vol. 2, pp. 467–472. IEEE (2016)
2. Al-Araji, Z.J., Ahmad, S.S.S., Farhood, H.M., Mutlag, A.A., Al-Khaldee, M.S.: Attack graph-based security metrics: concept, taxonomy, challenges and open issues. In: BIO Web of Conferences, vol. 97, p. 00085. EDP Sciences (2024)
3. Almutairi, O., Thomas, N.: Performance modelling of attack graphs. In: Forshaw, M., Gilly, K., Knottenbelt, W., Thomas, N. (eds.) PASM 2022. CCIS, vol. 1786, pp. 1–26. Springer, Cham (2023). https://doi.org/10.1007/978-3-031-44053-3_1
4. Cho, J.H., et al.: Toward proactive, adaptive defense: a survey on moving target defense. IEEE Commun. Surv. Tutor. **22**(1), 709–745 (2020)
5. Hermanns, H., Krämer, J., Krčál, J., Stoelinga, M.: The value of attack-defence diagrams. In: Piessens, F., Viganò, L. (eds.) POST 2016. LNCS, vol. 9635, pp. 163–185. Springer, Heidelberg (2016). https://doi.org/10.1007/978-3-662-49635-0_9
6. Hong, J.B., Chung, C.-J., Huang, D., Kim, D.S.: Scalable network intrusion detection and countermeasure selection in virtual network systems. In: Wang, G., Zomaya, A., Perez, G.M., Li, K. (eds.) ICA3PP 2015. LNCS, vol. 9532, pp. 582–592. Springer, Cham (2015). https://doi.org/10.1007/978-3-319-27161-3_53
7. Hong, J.B., Kim, D.S.: Assessing the effectiveness of moving target defenses using security models. IEEE Trans. Dependable Secure Comput. **13**(2), 163–177 (2015)
8. Ivanchenko, O.: Analytical and stochastic method in order to build safety and security block diagrams of cyber assets of scada system for critical infrastructure. Syst. Technol. **1**(57), 81–106 (2019)
9. Kao, Y.-C., Hwang, Y.-P., Wang, S.-C., Peng, S.-L.: An extension of attack trees. In: Peng, S.-L., Wang, S.-J., Balas, V.E., Zhao, M. (eds.) SICBS 2017. AISC, vol. 733, pp. 79–85. Springer, Cham (2018). https://doi.org/10.1007/978-3-319-76451-1_8
10. Kim, H., et al.: Time-based moving target defense using Bayesian attack graph analysis. IEEE Access **11**, 40511–40524 (2023)
11. Kumar, R., Stoelinga, M.: Quantitative security and safety analysis with attack-fault trees. In: 2017 IEEE 18th International Symposium on High Assurance Systems Engineering (HASE), pp. 25–32 (2017). https://doi.org/10.1109/HASE.2017.12

12. Lallie, H.S., Debattista, K., Bal, J.: Evaluating practitioner cyber-security attack graph configuration preferences. Comput. Secur. **79**, 117–131 (2018). https://doi.org/10.1016/j.cose.2018.08.005
13. Lallie, H.S., Debattista, K., Bal, J.: A review of attack graph and attack tree visual syntax in cyber security. Comput. Sci. Rev. **35**, 100219 (2020)
14. Landwehr, C.E.: Cybersecurity: from engineering to science. In: Developing a Blueprint for a Science of Cybersecurity, vol. 2 (2012)
15. Leversage, D.J., Byres, E.J.: Estimating a system's mean time-to-compromise. IEEE Secur. Priv. **6**(1), 52–60 (2008). https://doi.org/10.1109/MSP.2008.12
16. Lopuhaä-Zwakenberg, M., Stoelinga, M.: Attack time analysis in dynamic attack trees via integer linear programming. In: International Conference on Software Engineering and Formal Methods, pp. 165–183. Springer, Cham (2023)
17. Maglaras, L.: From mean time to failure to mean time to attack/compromise: incorporating reliability into cybersecurity. Computers **11**(11) (2022). https://doi.org/10.3390/computers11110159. https://www.mdpi.com/2073-431X/11/11/159
18. McQueen, M.A., Boyer, W.F., Flynn, M.A., Beitel, G.A.: Time-to-compromise model for cyber risk reduction estimation. In: Quality of Protection: Security Measurements and Metrics, pp. 49–64. Springer, Cham (2006)
19. Nagaraju, V., Fiondella, L., Wandji, T.: A survey of fault and attack tree modeling and analysis for cyber risk management. In: 2017 IEEE International Symposium on Technologies for Homeland Security (HST), pp. 1–6 (2017). https://doi.org/10.1109/THS.2017.7943455
20. Naik, N., Grace, P., Jenkins, P., Naik, K., Song, J.: An evaluation of potential attack surfaces based on attack tree modelling and risk matrix applied to self-sovereign identity. Comput. Secur. **120**, 102808 (2022)
21. Ortalo, R., Mehdi, C., Michel, P.Y., Girard, V.: Experimenting with quantitative evaluation of security based on attack graphs. In: Proceedings of the 3rd International Information Survivability Workshop (IISW 1999), pp. 134–151 (1999)
22. Pieters, W., Davarynejad, M.: Calculating adversarial risk from attack trees: control strength and probabilistic attackers. In: Garcia-Alfaro, J., Herrera-Joancomartí, J., Lupu, E., Posegga, J., Aldini, A., Martinelli, F., Suri, N. (eds.) DPM/QASA/SETOP -2014. LNCS, vol. 8872, pp. 201–215. Springer, Cham (2015). https://doi.org/10.1007/978-3-319-17016-9_13
23. Ponochovnyi, Y., Ivanchenko, O., Kharchenko, V., Udovyk, I., Baiev, E.: Models for cloud system availability assessment considering attacks on CDN and ML based parametrization. In: Proceedings of the 6th International Conference on Computational Linguistics and Intelligent Systems (COLINS 2022). CEUR Workshop Proceedings, vol. 3171, pp. 1149–1159. Gliwice, Poland (2022)
24. Rausand, M., Høyland, A.: System Reliability Theory: Models and Statistical Methods. Wiley, Hoboken (2004)
25. Schneier, B.: Attack trees. Dr. Dobb's J. **24**(12), 21–29 (1999)
26. Simpson, A., Dellago, M., Woods, D.: Formalizing attack trees to support economic analysis. Comput. J. **67**(1), 220–235 (2024). https://doi.org/10.1093/comjnl/bxac170
27. Xu, S.: Cybersecurity dynamics: a foundation for the science of cybersecurity. In: Proactive and Dynamic Network Defense, pp. 1–31 (2019)
28. Yoon, S., et al.: Moving target defense for in-vehicle software-defined networking: IP shuttling in network slicing with multiagent deep reinforcement learning. In: Artificial Intelligence and Machine Learning for Multi-Domain Operations Applications II, vol. 11413, pp. 617–626. SPIE (2020)

# Author Index

GPSR Compliance
The European Union's (EU) General Product Safety Regulation (GPSR) is a set
of rules that requires consumer products to be safe and our obligations to
ensure this.

If you have any concerns about our products, you can contact us on

ProductSafety@springernature.com

In case Publisher is established outside the EU, the EU authorized
representative is:

Springer Nature Customer Service Center GmbH
Europaplatz 3
69115 Heidelberg, Germany